D1302631

Mobile Media and Applications – From Concept to Cash

Mobile Media and Applications – From Concept to Cash

Successful Service Creation and Launch

Christoffer Andersson, *Ericsson, Spain*
Daniel Freeman, *Ericsson, Sweden*
Ian James, *Ericsson, South Africa*
Andy Johnston, *Ericsson, China*
Staffan Ljung, *Ericsson, Sweden*

John Wiley & Sons, Ltd

Published by   John Wiley & Sons Ltd, The Atrium, Southern Gate, Chichester,
West Sussex PO19 8SQ, England

Telephone (+44) 1243 779777

Email (for orders and customer service enquiries): cs-books@wiley.co.uk
Visit our Home Page on www.wiley.com

Other Wiley Editorial Offices

John Wiley & Sons Inc., 111 River Street, Hoboken, NJ 07030, USA

Jossey-Bass, 989 Market Street, San Francisco, CA 94103-1741, USA

Wiley-VCH Verlag GmbH, Boschstr. 12, D-69469 Weinheim, Germany

John Wiley & Sons Australia Ltd, 42 McDougall Street, Milton, Queensland 4064, Australia

John Wiley & Sons (Asia) Pte Ltd, 2 Clementi Loop #02-01, Jin Xing Distripark, Singapore 129809

John Wiley & Sons Canada Ltd, 22 Worcester Road, Etobicoke, Ontario, Canada M9W 1L1

Wiley also publishes its books in a variety of electronic formats. Some content that appears
in print may not be available in electronic books.

Library of Congress Cataloging-in-Publication Data:

Mobile media and applications, from concept to cash : successful service
creation and launch / Christoffer Andersson ... [et al.].
    p. cm.
Includes bibliographical references and index.
ISBN-13: 978-0-470-01747-0 (cloth : alk. paper)
ISBN-10: 0-470-01747-3 (cloth : alk. paper)
1. Cellular telephone systems. 2. Wireless communication systems.
3. Mobile communication systems. 4. Cellular telephone services industry.

TK5103.2.M58 2006
658.8'72–dc22

                            2005025166

British Library Cataloguing in Publication Data

A catalogue record for this book is available from the British Library

ISBN-13: 978-0-470-01747-0
ISBN-10: 0-470-01747-3

Contents

Acknowledgements

In undertaking this task we have relied on the expertise and experience of many talented individuals from a broad spectrum of disciplines. In creating a book to provide practical advice and tangible experiences we would like to pay strong tribute to our key contributors who have graciously given their time and shared valuable expertise to make this book something unique.

Thanks go to Charlotta Berg, Jonas Kåveby and Mattias Eriksson from Great Media who provided us with unique insights into the glamour industry. Charlotta provided valuable advice and experience in this area as well as on the many pitfalls and hurdles that need to be tackled. Regarding the finer details of content creation, Jonas and Mattias' advice and tips in this area are most appreciated.

Richard Schwartz, CEO of Solomio, has been one of the industry's drivers for enhanced voice services and has provided an excellent contribution on the importance of these areas as well as invaluable insights, based on real life experiences, into the successful deployment of services.

Much appreciation is owed to Thabiet Allie of MTN South Africa for his inspired marketing ideas and for taking the time to review chapter nine. Thanks also to Halima Khan of UIP South Africa.

A special thanks to Robert Henrysson, CEO of Jadestone, and Sven Hålling, CEO of Terraplay, for providing interesting perspectives on mobile gaming. Volker Hirsch of Mforma has also been of great help in giving further insights into this area. Robin Rutili, Head of Mobile Gambling at Mansion, has helped us to understand the practical issues for implementing complex betting and gambling services as well as their political and psychological effects. Kent Lundberg helped reduce our cognitive load.

In the TV and music areas, Roma Khanna, Senior Vice President Content at CHUM Television, has been an invaluable source of real experiences of mobilizing media. To this, Tapio Anttila, the mobile media guru of the Los Angeles area, added the Hollywood perspective. Additionally, the significant contribution of Christian Plenk must be mentioned for driving both Ericsson and the industry towards mobile media.

Oliver Holle, from 3United, has clearly had a big influence on our strive to understand media and content industries and his company has been a great partner and catalyst towards placing music and radio content into our mobiles in a personalized way. Similarly, we owe Kate Hess greatly for insights as well as collaborative experience in putting the right TV content into the right mobile context. Other prominent contributors in understanding industries include Erik Ott (Rainmaker Capital), Peter Laurin (Ericsson) and Miguel Rosado Boulet (industry expert and consultant in Spain).

Thanks go to Garret Molloy and Kirk Tang of MobileAware who provided wisdom on the art of rendering content to mobile devices. Special thanks also go to Vishant Vora, Chief Technical Officer & Vice-President, and Florin Geanta, Technology & Services Development Director, of Mobifon S.A for their expert and insightful views on the landscape of the operator service environment.

Thanks to Anoop Nathwani and Scott Allen at Surfkitchen for their input and support related to handset customization and mobile desktop software. Thanks also to O2 for allowing us to highlight their O2 Active service.

A picture says a thousand words. In the hands of a talented illustrator it says far more. We have benefited from the excellent skills of Claes-Göran Andersson who supported us and sacrificed his holiday to provide us with a series of excellent illustrations throughout the book – ranging from detailed diagrams to extremely amusing images, which have provided this book with significant value. Cover images were provided courtesy of Pär Altan (Trading Fate, www.tradingfate.com, concert image) and Mansion (poker cards).

As well as all the great people within Ericsson that build these mobile media and applications solutions every day, we particularly want to thank Kurt Sillén, VP Ericsson Mobility World, and Gunilla Fransson, VP Service Layer Solutions, for believing in us and this project. Kurt was the one pushing us to get started and this is just one of his contributions to building this emerging industry.

Special thanks are due to the many reviewers and contributors that have provided critical and valuable input and insights to this book over and above their own busy schedules.

Sebastian Lind, Urban Falk, Magnus Ekhed and Paolo Colella provided great input on the service layer and various integration aspects. Special thanks to Mark Campbell-Smith, Jose-Angel Hernandez-del-Pozo, Raymond Lee and Martin Thomas for their practical insights and experiences on deployments gathered in action from the trenches! The same goes for the Ericsson System Integration team in Spain/Portgual, including Magnus Rahm, Josep-Maria Balaguer, Ugo de Tommaso, Javier Cano, Guillaume Martin, Isac Palma, Antonio Fernandez and Ana Diaz. Thanks also to Michal Kubik for his valuable input and his trusted critical eye for detail, as well as to Roland Schlosser for his insights and illustrations on the treacherous nature of device and content verification.

Our esteemed colleagues at Ericsson Consumer and Enterprise Lab, Henrik Pålsson, Charlotte Karlberg, Renis Rahn and Susanna Lewis have been invaluable for their contribution to customer understanding, both within this book and in the wider telecoms world. Dr. Michael Björn's deep understanding of the Japanese market and its customers requires a special mention in this context.

Many thanks to Stefan Berggren, Martin Jönsson and Jan Johannesson of Ericsson Mobile Platforms, and Ove Persson, Björn Hallare and Karin Bering for giving valuable insights on the development of mobile devices. This also includes our friends at Sony Ericsson, most notably Ulf Wretling. We would also like to thank Lena Olsson, Ericsson Business Consulting, for sharing her deep knowledge in service development aspects.

Thanks go to Robert Skog for views into Single Sign On and more, to Jan Gabrielsson for the office & factory analogy, to Peter Arnby for his standards related insights, and to Ulf Olsson and Marc LeClerc, some of whose finely crafted words have been of inspiration here and there. Thanks also to Fadi Pharaon for enlightening on some of the ins, outs, tips and tricks in the mobile portal area.

Practicing what we preach, we have worked with the best possible marketing driver for this project, Maggie Curran.

Finally, authoring a book in your spare time inevitably takes time away from family and friends and creates an additional burden for those close by. Without the support, encouragement and dedication of these people, this book would never have been completed. Thanks and love for their patience and never-ending support therefore go to Anna, Lisa, Marta, Malin, Michelle and Ciaran.

Editor and Author

Christoffer Andersson, based in Madrid, is Director of Business Development Service Layer and System Integration for Ericsson (Spain and Portugal). He has eight years experience in applications development, product management and marketing of services for fixed and mobile networks.

He currently holds nine patents in this area, and is a frequent writer for various publications. His previous book, 'GPRS and 3G Wireless Applications – The Professional Developer's Guide', was published by John Wiley & Sons, May 2001.

Authors

Daniel Freeman has been driving the realization of mobile media and application solutions throughout the creation and launch of 2.5G and 3G networks for over seven years.

In this time, Daniel has promoted the importance of service assurance and verification activities around the design and launch of these new media services. He has recently also been working heavily in the areas of Music, Mobile TV and Gaming to ensure the availability of comprehensive solutions that meet our consumer's requirements for appealing, accessible and user-friendly services.

Ian James is a Senior Business Consultant with Ericsson. Ian has over 25 years of experience in the telecommunications and data communications industries, both in technical and commercial roles. Since January 2004 Ian has been based in Johannesburg, South Africa and advises Ericsson's customers in Southern Africa on their strategy for 2G, 2.5G and 3G data services, in particular the revenue opportunities from mobile content and applications. Ian spends much of his time working within mobile operators' marketing teams on projects such as portal launch programmes and content strategy development.

Andy Johnston is currently Chief Solution Architect for the Ericsson Service Layer & System Integration business in China. Andy is a software engineer who has spent the last 10 years in the trenches of the mobile telecommunications industry. Six of those years have been spent in the effort to design and deliver solutions for a useful Mobile Internet.

 **Staffan Ljung** is Principal Consultant within Ericsson's Global Services organization. From his current base in Prague he manages a team of business advisors in Market Unit Central Europe focused on leveraging new technologies to enhance the service offering of its clients.

Staffan has a long track record of successful projects from various telecom assignments around the globe, focused on the launch of mobile services.

Introduction

OUR GOALS WITH THE BOOK

Our ambition with this book is to offer a fresh and wide perspective on a dynamic industry by drawing together data, information and experience, and synthesizing these into useful technological and business conclusions. We decided to ground our approach in combining market and customer understanding gained in our personal experiences and observations as practitioners, with the implications of those observations on networks, devices, operations and so on. In doing so we chose to emphasize:

- a preference for practical aspects based on real-life cases, rather than a formal or scientific approach;
- striking a balance between business challenges and technical implementation in an accessible format;
- looking at the industry from three angles – media, IT and telecom.

Naturally we envisioned ourselves adding in a little humor here and a little literary or graphical flourish there, although we are certainly far from competent in these areas!

OVERVIEW

This book is about making mobile media and applications happen, such as gaming and mobile TV, both from a technical and a business perspective. This is neither a programming book nor does it try to define new theoretical models for marketing. Instead, it is a complement to such books by looking at real case studies, giving practical advice to learn from the successes and failures of others.

We start this by giving you a strong understanding in all areas:

- Understanding the customer – what drives the customer to play games or send pictures?
- Understanding the market – for example how many devices capable of running this application are out there?
- Understanding how to make things to work together, e.g. creating a TV stream encoded for mobile devices and implementing charging per download rather than per kilobyte.

OUTLINE OF THE BOOK CHAPTERS

This book takes the innovative and groundbreaking step of presenting its message in a number of chapters! Although the chapters can be read separately and in any order you desire, certain threads flow through the entire book and there is a benefit to be accrued from starting at the beginning and following the time honored, beaten path through the book, chapter by chapter. For example, we attempt to create an understanding of the customer early in the book, which we build upon in later chapters. Anyway, it's your book now, so we will stop interfering!

Chapter 1, 'Fundamentals of the Industries', introduces the players from the IT, telecom and media industries and describes the value chains and working relationships. This highlights the very different viewpoints and how many existing approaches have failed, while there are some successes that we can learn from.

Chapter 2, 'Understanding the Customer', starts from the user perspective, both consumer and enterprise, and considers why people behave like they do. We introduce segmentation models that divide people and businesses into various categories and look at the underlying reasons why some applications have been successful and others have not.

Chapter 3, 'Creating a Winning Service Offering', explains how operators build service roadmaps, media companies evolve their offerings by adding interactivity, and developers try to collaborate with them both to get their applications on the market. The chapter takes a holistic view of a whole service offering and leverages synergies between the different services and channels.

Chapter 4, 'Designing Services', dives into a technical classification of services and the far-reaching implications of early design choices. This is followed by a round-trip of design guidelines and tips needed to ensure a superior user experience and mobile optimized end-to-end performance. The chapter also covers the fundamentals of mobile media creation, looking at mobile images, video and audio media types.

Chapter 5, 'Managing the Customer Experience', considers how to attract and retain the customers with a rich and dynamic experience. This includes ways of increasing the value of the service offering as well as how to remove the many barriers that still prevent customers using mobile services.

Chapter 6, 'Mobile Devices – Leading the Way', goes through the evolution of devices, from when they followed the networks to today, when applications start from the device. Furthermore, it describes how the devices work 'under the hood' and their impact on service development. This includes not only the technical aspects but also the economical and strategic impact of device penetration, features and capabilities on service take-up.

Chapter 7, 'Service Environment', introduces and deals with selected functions, technologies and architecture of the machine room that enable delivery and allow management of services. Focus is placed on how technologies can and should cater to business needs and business viability as well as how new technological features and qualities can inspire new services or new and better ways of offering existing services.

Chapter 8, 'Deployment of Services', focuses on the practical aspects of the deployment and integration of complete services. We cover many practical aspects such as the importance of careful preparation and scoping, the need for strict project and quality control as well as numerous technical details based on real-life experiences from previous deployments to help avoid some of the obvious and not-so-obvious pitfalls.

Chapter 9, 'Commercial Launch Experiences', follows up on the previous chapter of deploying services, but approaches it from an outside-in perspective. The classic marketing themes of pricing, packaging and promotion are discussed, based on case studies of real-life launches.

Chapter 10, 'Feedback and Quality', closes the circle by introducing feedback loops and methodologies for keeping the service offering alive and attractive. This includes monitoring and assurance of quality and usability to continuously improve the customer experience.

WHO SHOULD READ THIS BOOK?

Although we hope that people of various backgrounds and interests may read this book, our intended readers are decision-makers and other professionals in the IT, telecom or media industries; people whose everyday efforts are key to the success of this industry. We hope our use of case studies and concrete experiences has something to offer for a wider audience.

COMPANION WEBSITE CONTENT

The companion website can be found at www.mobilemediaapplications.com and has many similarities with the extra material nowadays frequently attached to a blockbuster movie DVD. The website contains links to useful information as well as material and case studies that we could not fit into the book. The Directors' cut of the book itself, however, you are holding in your hand.

1

Fundamentals of the Industries

1.1 MOBILE SERVICES – FIASCO OR ROARING SUCCESS

The mobile Internet industry is an enigma to many. On one hand there is an amazing emotional dependence of billions of people on having their mobile phone close by and their increasing use of it for communication, messaging, personal organization and entertainment. At the same time the industry has seen tremendous financial setbacks and a sluggish uptake of new multimedia services. On one hand the pessimists are claiming that a success like Vodafone Live does not generate more than a few percent of the company's revenues (see Figure 1.1 for a generic operator revenue split) and most data revenues come from simple text messaging, SMS (Short Message Service). On the other hand the optimists show how revenues from simple ringtones already in 2003 formed 10 % of the, in other aspects troubled, US$30 billion plus music industry.

Clearly the telecom industry has made mistakes in creating massive hype around the next big thing, while sometimes in the process completely missing the successes that happen around us. While we saw TV commercials showing spacemen on surfboards boasting the amazing potential of WAP (Wireless Application Protocol, a technology which we will see is not as bad as is often claimed), text messaging (SMS) was skyrocketing and teenagers were carefully selecting the right ringtones and wallpapers to show they were cool and hip.

1.2 WHO NEEDS MOBILE SERVICES, REALLY?

Well, it would be somewhat of a spoiler if we already in Chapter 1 revealed all our findings on who needs mobile services and why, wouldn't it? First of all, we have to be humble about the nature of humans and our capability to change our behavior. It takes time to change behavior, lots of time. Probably, the mobile telecom industry has been spoiled by the incredible rate of take-up for mobile telephony and the Internet, as seen in Figure 1.2.

Mobile Media and Applications – From Concept to Cash: Successful Service Creation and Launch Christoffer Andersson, Daniel Freeman, Ian James, Andy Johnston, Staffan Ljung
© 2006 John Wiley & Sons, Ltd

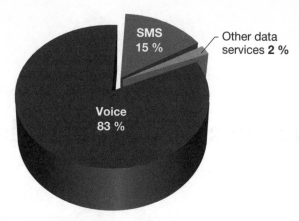

Figure 1.1 Revenue split for an example European operator.

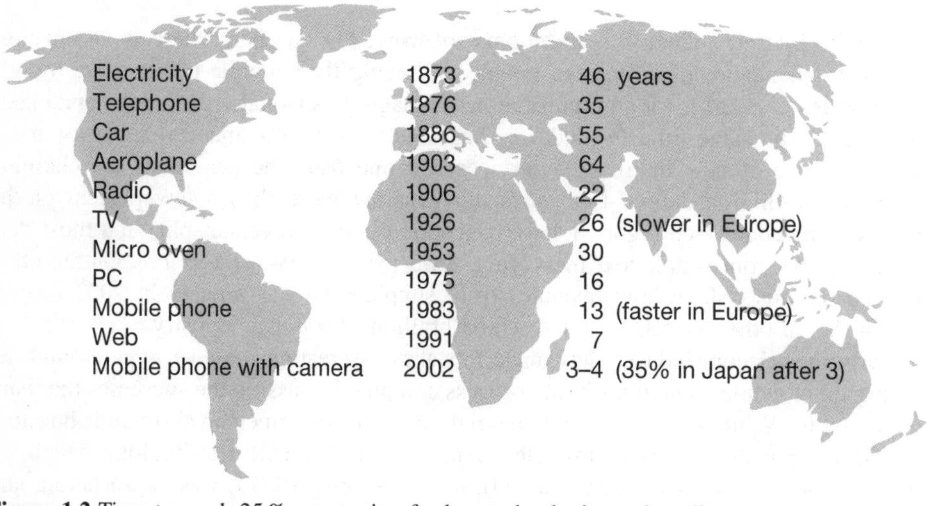

Electricity	1873	46	years
Telephone	1876	35	
Car	1886	55	
Aeroplane	1903	64	
Radio	1906	22	
TV	1926	26	(slower in Europe)
Micro oven	1953	30	
PC	1975	16	
Mobile phone	1983	13	(faster in Europe)
Web	1991	7	
Mobile phone with camera	2002	3–4	(35% in Japan after 3)

Figure 1.2 Time to reach 25 % penetration for key technologies and appliances.

Fifteen years ago most of us knew the concept as car phones rather than mobile phones, big bulky things that maybe in the future would be interesting to businessmen if the size decreased. In 2005 there are around 2 billion people using mobile phones[1] around the world and over 680 million phones (257 million camera phones,[2] compared with 74 million digital cameras[3]) have been sold in one year. Clearly, humans have a very strong need for mobility and communication and the mobile is becoming one of the few things you do not want to leave home without.

---

[1] Source: Ericsson.
[2] Source: ZDNet IT facts site, http://blogs.zdnet.com/ITFacts/.
[3] Source: IDC.

In this context we must remember that different markets are in different stages of their development. On one hand we see rapid subscriber growth in Africa and parts of Asia (4–5 million new subscribers every month in China), but on the other hand some Asian and European countries have reached close to 100 % penetration or even more. In those markets, the mobile operators are entering a new phase where not only the number of new subscribers is low but sometimes also the average revenue per user (ARPU) is declining (despite higher usage). It is quite easy to see that such operators have to look for new sources of revenues to please their shareholders. Therefore, it is not only customer need driving the development of mobile media and applications.

Before we start looking into the industries involved in the chase for the next big thing, the next killer application, one might ask if there really is any money in mobile services? Isn't this only about fun things for teenagers killing time? While many small companies make their living solely from this area, others are more skeptical. It is a certainly a tough challenge to understand what consumers want and what they are willing to pay for. For now, we will only note that consumers are paying €1–4 for a simple ringtone and €8/month for mobile TV that only has one to two picture updates (frames) per second, and hope that is enough incentive for you to keep reading and discovering.

1.3   THE TELECOM, MEDIA AND IT INDUSTRIES COMING TOGETHER

Getting complex services to fit customer needs and behavior is in itself a difficult task. With several industries coming together, it becomes a daunting challenge. Telecom sees services beyond voice as the future for maintaining growth and profitability, while markets are closing in on 100 % penetration and voice prices drop. Media sees the possibilities of the huge customer base, richer devices and working payment models as means to expand its core business, while IT sees a natural evolution of cutting the cord to the Internet, see Figure 1.3.

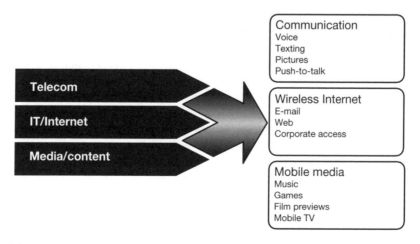

Figure 1.3 Industries converging.

This might sound like start of a happy three-way marriage (which would be a paradox in itself). However, when industries come together and start collaborating, it is key to understand where the different parties come from and what they bring with them.

1.3.1 Telecom – a short story in time

In the early 1990s the second generation (2G) digital mobile systems, like GSM (Global System for Mobile communications) were introduced, which all introduced higher capacity, smaller phones and better talk and standby times than the old analog systems. While many today take the 2G successes for granted, there were several troubles to begin with.

In 1992 a business magazine[4] wrote:

[...] there is a complete mess around the new European mobile system GSM. The utterly important launch of a new and expensive system is mostly resembling a fiasco and GSM is delayed a year.

In 1994, two years later, another respected newspaper[5] said that

[...] while mobile telephony is growing quickly almost all customers choose analog phones over the GSM phones, mostly due to superior coverage.

As 2G was rolled out during the 1990s the only service was voice, with SMS text messaging added as a mechanism to alert customers when new voicemails had arrived. When first Nokia and then other devices manufacturers during late 1990s started promoting ring-tone downloads via SMS, a new range of services started to emerge. This was riding on the need of customers to personalize their devices and in parallel the person-to-person SMS traffic grew quickly as the devices became easier to use and screens got bigger. Interestingly, however, the telecom industry has always dreamed about the 'next big thing', like 3G and IMS (IP multimedia subsystem), rather than evolving the current successes.

As an example, already in 2000, 500 million SMSs were sent every day around the world (rising to over 1 billion per day in December 2002). Around that time, everyone was busy hyping 3G and driving the prices for its spectrum (frequencies) to stellar heights. With most operators buying expensive spectrum and the much-anticipated WAP services failing to impress, the industry went into depression for a couple of years, coming back around 2004.

These miserable times still plague the mobile services industry and lower its willingness to take chances and try new things. While this might sound dampening on anyone's enthusiasm, the mobile telecom industry has been able to reach an impressive number of customers, and these customers not only use services like SMS, ringtone and game downloads, but they are also prepared to pay for those services. Consequently, the

---

[4] Source: not disclosed in order not to point fingers at a particular magazine. This is just one example of the overall media view.

[5] Source: not disclosed, for the same reasons.

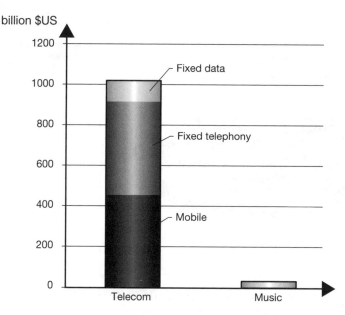

Figure 1.4 Comparing the sizes of the telecom and music industries[6].

telecom industry today is enormous compared with other industries, for example music (Figure 1.4).

1.3.1.1 Japan and Korea leading the way

In Japan, the market development has been ahead of the rest of the world and, while one operator was the initial leader, the other two then pioneered some of the following steps. NTT DoCoMo, the market leader, in 1997 tried offering value-added mobile services to business customers, but did not succeed. In mid-1997, DoCoMo changed their vision and created a consumer-oriented offering: the *i-mode* concept. i-Mode was launched in February 1999, built around a micropayment system (paying via the phonebill) and connecting to external service providers (Disney, Bandai, etc.) that provided content (mostly simple entertainment, like wallpapers and ringtones). The win–win business models (91 % of the content revenues for the content companies), consumer focus, application-centered devices and simple entertainment focus created a rapidly growing service offering. While the Western countries were busy trying to create a wireless Internet, Japan was heading more towards mobile media (Figure 1.5).

While the success in Japan has frequently been published and analyzed, South Korea's contribution to the mobile media and applications evolution is seldom explored. South Korea is actually a more applicable role model than Japan, having high Internet penetration using the same technologies as in Europe and the US. We will look further into the success factors of South Korea in Chapter 9.

---

[6] Source: mobile-Ericsson 2004 estimate, fixed from 2002, music from ifpi 2002.

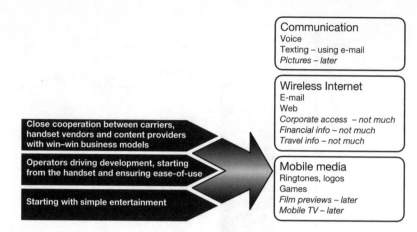

Figure 1.5 Initial mobile services in Japan.

1.3.1.2   3G and mobile services growing around the world

As 3G finally got started in the rest of the world in 2003/2004, it initially suffered from the same difficulties of insufficient number of devices and 3G coverage as in Japan. Now the challenge was to find the magic formula and service packages that would attract the customers. While many thought it would be video, it turned out to be the old champion voice – cheap voice.

Hutchinson, with the mobile operator brand 3, has, together with DoCoMo, been one of the strongest promoters of 3G services and networks. Initially, 3's growth was sluggish and customers did not change operator just to be able to use new 3G services, like video calls. When 3 changed strategy to also compete on price for voice calls, often free within 3's networks and to a fixed monthly fee (flat rate) for other calls, its growth accelerated quickly. With the number of subscribers increasing, 3 keeps focusing on new services beyond voice (like video, games and music) and in this book we will describe several examples.

Also, outside Japan the need for superior quality and coverage has been key for 3G growth, with customers reluctant to settle for less than they had yesterday. A mistake often made in the beginning was to build out the 3G network by trying to maximize the number of people covered. While this sounds logical at first glance, the result is often coverage in all cities but very little coverage outside the main urban areas. The customers quickly saw that 3G did not work when they were in their car on their way to work or at the beach. Consequently, a wiser strategy has been to ensure surface coverage first, ensuring that the key regions are quickly built out with complete coverage.

1.3.2   Convergence

Convergence is a popular topic, but what is it really? We have talked about the convergence of industries but we also see the convergence of technologies and services. Normally, the following three areas are used to describe convergence:

- *convergence in services*, e.g. playing Tetris on your TV against someone who uses a mobile phone or Game Boy;

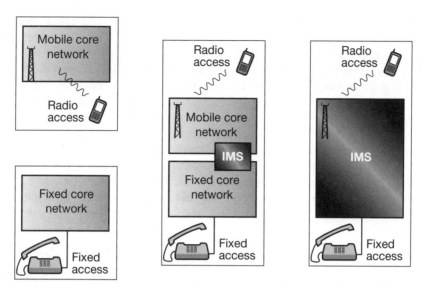

Figure 1.6 Evolution towards one common core network for fixed and mobile operators.

- *convergence in devices*, e.g. laptops or phones with wireless LAN (local area network) for both data and voice usage; an important concept is what we call Always Best Connected (ABC), meaning that in the future the device should automatically find and use the best network for each situation;
- *convergence in networks*, e.g. having different access networks but the same Internet protocol (IP)-based core networks for both mobile and fixed networks.

In this book we will mainly focus on services and for certain case studies mention the convergence aspects. We have seen more and more services benefiting from letting the customer recognize the actual service and content regardless of whether they are using the PC, TV or mobile. The network convergence aspects will also be touched upon, as these have an impact on how you build the infrastructure and architecture needed to launch services. The IMS is a fundamental part of this evolution, installed by both fixed and mobile operators in order to obtain the same core network infrastructure (Figure 1.6).

1.3.3 The IT and Internet story – going mobile

Discussion of the mobile industries easily becomes focused on Japan and Europe; the USA takes a much more prominent role when it comes to the Internet and the IT industries around it. While we won't go into the stories of Moore's law of rapidly increased computer power per chip or how Microsoft put Windows PCs on everyone's desktop, we will focus on how Internet is going mobile.

The IT industry has long been advocating openness, but in a different way than telecom. Standards are important, like IP, TCP and Web Services, but also openness on a deeper level. For starters, the client hardware (for normal people, the home PC), is built on components that amateurs can put together themselves. While most home PCs have Microsoft

Windows as their operating system, a free alternative is available with Linux. Home users choose Linux less frequently than businesses, which use it as a server environment and appreciate its openness and possibility for customization.

As wireless LANs and 3G PC cards have made their entry, it is beyond doubt that the IT world is going wireless even from the customer perspective. Additionally, customers expect to access the same (or similar) information and applications when on the move as from their desktops. This makes it natural to also utilize concepts and design principles from the IT world when designing mobile services. In this reasoning, however, it is important to also remember some of the key characteristics that have made mobile phones the companion of over 1 billion people after around 10 years. Ease of use and reliability have always been trademarks (not that this is fulfilled for all service offerings today), and customers have very little tolerance of a hand-held device that has to be restarted every once in a while.

1.3.3.1   (Almost) Everything is free on the internet, including voice calls

In the early days of the IT bubble, billions were spent on creating companies that (later, sometime) would make their revenues mainly from advertising and therefore offered many things for free. Unlike printed press and books, information on the Internet is mostly free and looks likely to remain that way. While this might also reflect the perceived value by customers, it has led to difficulties in charging for digital products, whether it be software or music. The mobile channel, in contrast, has long offered possibilities of letting customers pay via their phone bill or prepaid card.

From the start of the Internet, experiments have been made around enabling telephony (voice-over-IP) of these networks. After a number of basic voice-over-IP services with unimpressive quality, something happened in August 2003 that also would impact the telecom world. The peer-to-peer[7] service Skype was launched and started to gain customer acceptance. Its success was built around better quality than previous free voice-over-IP solutions as well as the easy possibility of also calling customers on their regular phones. In 5 min, a customer can download Skype, deposit €10 in an account and start calling his Skype friends for free and most other customers who only have regular phones for as little as €0.0017/r min. While many telecom operators had already started moving into voice-over-IP solutions, clearly Skype has accelerated this development. All has not been perfect with Skype, and the quality of service is still not on a par with regular fixed or mobile calls unless the customer only runs Skype and no other downloads/uploads. This comes from the lack of the ability on basic Internet networks to guarantee the speed, and you get what you get (best-effort). Fixed and mobile phone networks, on the other hand, have always been built to set aside specific channels per voice call to ensure constant quality. However, as bandwidths increase and more customers get 10 Mbit+ connections, this becomes less of an issue.

When will this impact the mobile industry? It seems that it will first impact the fixed telecom services, especially for calls between countries, as those are charged the same as national calls. However, as mobile systems get higher capacity and higher power

---

[7] Peer-to-peer, P2P, systems have most of the intelligence in the client, like a PC, rather than in the network.

devices with open operating systems, this and similar systems will be a consideration for all telecom players. Consequently, the convergent systems, like IMS mentioned above, provide voice-over-IP services by using Internet technologies but adapted to the strict requirements of voice. There are several regulatory considerations for voice-over-IP that are not completely developed, like ensuring emergency calls.

1.3.4 Mobilizing media and content

Media and content industries are eyeing the mobile channels for several reasons, and this area is evolving quickly. We see three main drivers for this:

- the drive to go digital, and the world becoming connected;
- the need for interactivity;
- the need for personalization and marketing.

First of all, many record labels, movie studios, TV channels and others see the global trend towards digitalization and the need to adapt their business models and offerings. The challenge is to meet customer needs for getting media content via the Internet and to earn direct and indirect revenues from this, in the form of ringtones, videoclips and merchandising.

While music went digital with the advent of CD and movies with DVDs, an even larger shift came when everything got connected via the Internet. After all, the main idea with storing music on LPs and later CDs was to facilitate distribution and packaging related to the artist. While CD singles were available, the cost for producing a disk for one or a few songs made the price high. With mp3 files becoming widely available during the late 1990s, customers saw both that songs could be downloaded for free and that music did not necessarily need to be distributed via whole albums. The whole music industry and later movie industry has been hurt by the subsequent file sharing (Kazaa, Direct Connect, BitTorrent), as well as the lack of good solutions for legal downloads. For several years (late 1990s, early 2000s), little happened in the area of legal solutions, prices remained high on CDs and the online file sharing of music rose quickly.

In 2003 this market changed, with the success of Apple's iPod portable music player and associated iTunes Music store, where songs could be purchased online for only $0.99 in the US and €0.99 in Europe (£1.14 in the UK). The iPod music player device was actually launched in 2001 but only took off in a big way when iTunes was introduced. This is clearly an example of how new customer behavior only happens when the whole end-to-end chain is ensured, in this case with the music software and online music library together with the portable music player. Other key success factors included customers being able to try out 30 s clips before buying songs and the ease of use. The revenues for the music download mostly goes to the music industry, the copyright holders, rather than Apple.

Looking at these media and content industries and their entrance into new (digital) channels, there are several aspects that these industries have in common. One of the most apparent features is the need for means to ensure copyright protection. This is not only a technical issue, with the need for digital rights management systems (DRM) in devices and networks, but also includes legal and commercial processes and systems. The main reason why the music industry has been afraid to explore the Internet is its traditional lack of payment mechanisms and means for protecting copyright.

The second prominent driver for mobilizing media, interactivity, represents an element that has been eagerly awaited by many media players. Many of the traditional media channels, like TV, CDs and radio, have primarily been broadcast media (for receiving information). The mobile, on the other hand, also lets customers send information and therefore constitutes an excellent tool for interactivity. This includes voting in a TV show or sending in comments, which now has become commonplace in many countries.

Finally, the mobile channel is starting to advance as a marketing tool, a difficult and sensitive area but with high rewards. One example was Volvo cars creating a marketing campaign around a mysterious story (The Mystery of Dalarö) that gradually evolved via interaction in papers, TV, Internet and mobile. Another example was Coca Cola's bottle hunt in Germany, where customers could find a code and a phone number on the bottles. From there the customer could call and find out what prize he/she had won and everyone won something, from ring-tones to multimedia messaging service (MMS) and funny calls that could be sent to friends. In this case, Coca Cola was not that interested in the direct revenues; the big gain was the 0.1–0.2% gains in beverage sales that they registered, which translates into a lot of money. This particular case is described in more detail in Chapter 9.

Frequently people from the outside see all media and content industries as one homogeneous industry, whereas those working in the mobile services value chain need to understand some key differences. While there are many different media and types of players, a good start is to understand the difference between media and content players, as illustrated in Table 1.1.

Example media companies include TV channels (MTV, Discovery Channel, Paramount Comedy) and newspapers, while content players are those producing, for example, TV programs, games (Sega, Electronic Arts), movies (Paramount, Lucasfilms) or news feeds (Reuters). Frequently, there exist as many differences (and conflicts) in opinion between media and content players as these do between either of them and mobile operators. For example, a content company like a movie studio often has a business model based upon upfront fees for the rights to use its content and brands. Here operators and media companies have more in common, with their direct access to the consumers and therefore a higher belief in sharing revenues based on usage growth.

Mobilizing media and bringing it to customers of mobile devices is often done in a stepwise approach, where the evolution starts with simple things, getting feedback and then gradually introducing new features. In the next section we will take a look at some different industries and how they are going mobile. First, however, we can summarize the key requirements on the mobile channel from media and content industries:

- distribution capabilities of mobile network that protect copyrights – still controlling the content!
- broadcast related business models and flexible billing services;
- interoperability and flexibility – easy content creation for all devices and ways to keep the content dynamic and changing

Table 1.1 Media companies vs content companies

	Media	Content
Main business	Advertising	Content production and distribution
Main mobile business	Mobile marketing Mobile backchannel to TV/print Mobile content distribution	Mobile content distribution through carrier portals through media portals
Attitude towards the mobile channel	Sell as many content and premium applications as possible to the existing customer base without annoying them	Sell its own content through as many channels as possible without losing protection of copyright
Digital rights management	Not important, more seen as 'necessary evil'. Community and viral effects are more important	Extremely important, very restrictive
Relation to end customers	Have broad, loyal customer base that they can sell into	Have no direct relationship with customer base
Relation to operators	Use carriers mainly for interconnection .	Often have long-standing relationships with carriers as content suppliers

A common theme among these players is that they want to control the customer experience and the way their content is consumed. This today is a very big challenge for the mobile channel, and the cause of many conflicts between media/content players and operators.

1.4 CONTENT AND MEDIA INDUSTRIES GOING MOBILE

Now let us take a slightly deeper look into some of the content and media industries that have their own traits. The concepts of convergence are also very visible in this context, as gaming, music and TV companies only see mobile services as another channel and not the center of their universe (as we in telecom sometimes seem to believe). We will also look into what each area is looking for in the mobile channel.

1.4.1 The music industry fighting for its rights

The ringtone industry has evolved in ways that have surprised many, and its revenues are significant, US\$3–4 billion, depending on which source you listen to. It is not easy to measure this because many players are involved, not just operators and record labels, but also smaller players that partly go outside the operators and/or labels. For example, a homemade ringtone, like the Crazy Frog (sounds like a frog that imitates a motorcycle),[8] does not use any artist or label that needs to be paid royalties. Crazy Frog is an example of how far the ringtones have penetrated our culture, being so popular that in 2005 it was released as a CD single! If that was not enough, the Crazy Frog remix in June 2005 reached the number 1 spot on the UK singles chart, beating all established artists.

Furthermore, some of these use newspapers, web sites or teletext to market their tones and thereby partly change the value chain. In these cases it is not the operator that is the brand towards the customer and the one doing the marketing, which could potentially be a challenge for operators. Perhaps this is natural model, as we will see when looking at the business models later in this chapter.

The step to polyphonic ringtones meant that phones could play more than one tone at a time, for example using the MIDI format that is common for electronic keyboards, while RealTones brought the sound of a real artist to the ringtone, e.g. using the mp3 format. Ringbacktones are used for a separate service, letting you hear a song or other recording when calling someone and waiting for an answer, replacing or complementing the usual beep-tones (e.g. hearing 'Wish you were here' from Pink Floyd while waiting for your friend to answer). Finally, many phones are today able to function as music players for audio streaming or download, which is described more in Chapter 6. Figure 1.7 not only illustrates the ringtone evolution but also shows how content has changed from interpretations (beep–beep) of the real material version that imitate a real song, to using the actual recording of an artist (shown towards the right). This of course changes the customer experience but it also changes the rights situation and the business models, as shown below.

For the monophonic and polyphonic ringtones, the ringtone provider only needs to clear the rights for the composer, while the mechanical rights for content arrangements belong to whoever created the ringtone in that format, usually the provider/aggregator. For

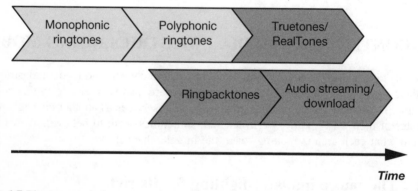

Time

Figure 1.7 Ringtone evolution.

---

[8] The Crazy frog animation, while promoted by content distributor Jamba! GmbH, was created by Erik Werquist with sound by Daniel Malmedahl (i.e. not a big-name record label).

Monophonic/polyphonic ringtones

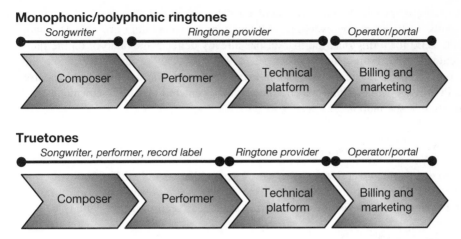

Figure 1.8 Content rights for monophonic and polyphonic ringtones.

example, U2 owns the content composer rights to its song Vertigo, while neither U2 nor its record company owns the mechanical rights for a polyphonic ringtone interpretation (owned by whoever does the interpretation). This led to frustration for the record labels, which were left out of the value chain (shown in Figure 1.8).

In truetones, the original artists and the record labels have the right to the performance and thereby receive a piece of the pie. This has, however, led to ringtone companies and aggregators getting squeezed in the middle between operators and record companies, who each take as much as 30–40 % of the revenue share. This is the reason why some aggregators and ringtone companies today use alternative artists to perform the latest hits (cover versions), thereby avoiding paying for the performance to the record label and original artist. Another alternative, as we saw earlier, is to use frogs to perform the latest hits.

1.4.1.1 Clearing of rights

Once someone wants to distribute ringtones or music tracks, the rights have to be cleared, paying the rights holders, which has not always been easy. The starting point is the way radio stations pay a royalty to an independent body (usually country specific). This body then distributes the royalties to the record labels/songwriters. With the mobile channel, however, the effects of digitalization are taken to their extreme. If you buy a ringtone in Sweden, from a Swedish operator and a Swedish ringtone company that has its server on Swedish soil, it is quite easy to see that the provider should pay the Swedish body, STIM, that handles rights for songs played on the radio. Now imagine that the ringtone company is still Swedish but the customer is German on holiday in Greece. In addition, the ringtone company has chosen to put its servers in Spain. Now it is not that obvious where the rights should be cleared anymore, although the industry is slowly coming to grips with these issues, at least for polyphonic ringtones. Real music clips represent a large challenge in clearing rights and some aggregators have been surprised to find out that acquiring a music clip for use as a truetone ringtone does not automatically give the right to use it as ringbacktone.

Coming up with a viable and sound business model for download of mobile music presents a challenge for everyone. Although mobile operators are able to charge US $2.00–5.00 for a ringtone, the market has already defined the value of online music file downloads (like iTunes and Napster) at $0.99 or €0.99. We are likely to see different flavors of mobile music services, some using the iTunes/Napster model of synchronizing the PC library while others will be more phone-centered and offer downloads of songs directly into the mobile.

1.4.1.2 More than just music

As the mobile music services evolve, we are seeing that not only different quality ringtones and music downloads are the drivers. One area is the radio services already mentioned, but TV and video are also growing in importance, as seen in the case study below.

Case study: 3 UK music service
In August 2004, British operator 3 enhanced its music offering with music videos, a service called Video Jukebox. Customers could choose between paying £1.50 per video (streaming, called QuickPlay, or download) or £10 for unlimited downloads. The music videos are offered via the mobile channel 6 weeks ahead of the single release and therefore serve as a marketing tool, in addition to the direct revenue. During the first 6 months, this service was accessed 10 million times at a time when 3 had 2.5 million subscribers in the UK.

This success has been built around launching the right content to the right audience at the right time (when a critical mass of capable devices was available). In the UK the TV penetration is around 95 %, with customers spending 50 times more time in front of the TV than using the mobile phone.[9] With a correctly positioned offering to this audience (putting content in the right context by relating it to favorite programs on the TV), focusing on young and entertainment-savvy customers, 3 reaped the benefits.

1.4.2   Radio interaction

Radio remains a medium that gets lots of time from customers, being available everywhere and not requiring much concentration, e.g. customers listen while driving a car. It is a business that has not changed much during the last 20 years or so, when commercial, advertising-financed channels emerged. In this book we focus on the commercial radio stations, whose core business is to understand and please the listeners in order to generate revenues via advertising (some public service stations still get their revenues from tax or other state funding). Running a radio station can in a simplified view be described as three branches, as shown in Figure 1.9. Looking at each of these three areas, we can identify the key needs that can form the basis for creating mobile services:

---

[9] World's Top-10 Mobile Services, analysis January 2005.

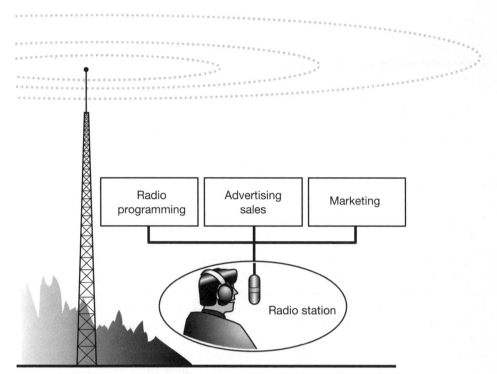

Figure 1.9 Key areas for a commercial radio station.

Firstly, the radio programming is the content, which most of the time is broadcast live, thereby differentiating from other media. The talkshow hosts and DJs are key to differentiation and success. By creating appealing and engaging content, listeners stay longer and advertisers are happy. Competitions and other formats where customers can interact are also growing in importance to achieve these goals.

Secondly, advertising sales is the branch that generates money from the above-mentioned programming. The advertisers want to see why the listeners of this particular station are best suited as a target group for its products and also that the brand images match. For example, a skiing resort with great party facilities looks for a radio station that is positioned as cool and youth-oriented, with the latest dance and party hits.

Finally, the marketing side supports the advertising sales by understanding the audience and creating programs that add value for the listeners. They use independent bodies for listener polls that show the ratings. Radio stations are quite sophisticated in mapping and understanding the target audience and its demographics. As mentioned above, this is essential for the companies that buy advertising.

With all this focus on generating money from advertising, is there any interest in using the mobile channel? There is. First of all, there are a number of services that directly add value for the advertising sales and marketing processes outlined above. Additionally, the direct revenues at times have also proven to be significant. Radio programming is often built around a large proportion of live broadcast shows, which increases the value of interaction.

In order to look into the creation of services that enhance radio, we show a success story from Austria.

Case study: mobilizing radio stations[10]

Ö3 is the largest radio station in Austria with around 50 % market share and 2.9 million listeners daily. It is part of larger group containing 11 radio stations nationwide (a common scenario for this kind of radio station). The focus for this particular station is hit music.

Ö3 specified the following objectives for mobile services:

- to increase the listener base and frequency through value-added services;
- to increase advertising sales by providing higher value promotion and rich demographic information;
- to build a highly targeted customer relationship management (CRM) database asset and generate direct revenue from the services.

The solution contained several steps. The first one was a simple service where customers would send an SMS (price €0.30) to a short number in order to receive information about the artist and song title of the song currently playing on the radio. When receiving this info, the customers could also easily access any associated product available, like ringtones, wallpapers or CD. Based on the information gathered, a customer profile is built up in the CRM database. Customers can then take advantage of this profile and use more services, like the *hit-reminder*. The hit-reminder allows listeners to sign up (via WAP, web or SMS) to receive an alert when a song is played next, see Figure 1.10.

The results are very promising and have led to further developments. The services have become very popular, early on having 150 000 customers generating 250 000 requests/

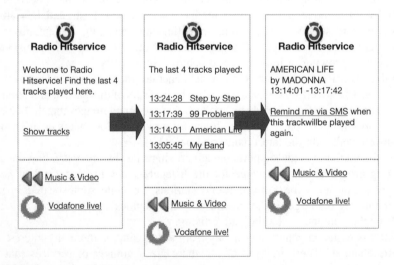

Figure 1.10 Radio Hitservice example screenshots.

---

[10] Source: 3United.

month. Out of these over 80 000 (more than 50 %) signed up as 'mobile club' members (i.e. available in the CRM database), a good indicator of customer loyalty. Additionally, the number of listeners grew to the highest level ever and advertising revenues increased.

1.4.3 The gaming industry putting the fun into our hands

The size of the gaming industry is hard to specify exactly, but it is larger than, for instance, the movie industry (and growing more quickly). Here we look at two main areas of this industry that are moving into mobile gaming. First of all we will examine the PC and console game industries. These are normally treated very differently, but for now we will look at them with those that work with very big screens and (increasingly) broadband network connections. Secondly, we have the small portable games that started in the 1980s with Nintendo and Sharp's development of Game-and-Watch and accelerated with the first Nintendo Game Boy in 1988.

During the later part of the first Game Boy's development, it was obvious that this portable format required a different kind of game to take off. The hardware was quite simple and the screen black and white, so the games had to be easy to use and addictive to compensate. The solution came along when the developers crossed paths with Russian mathematician and programmer Alexey Pajitnov. Pajitnov had in 1984 created a simple addictive puzzle game in his spare time, and this game had increased in popularity throughout Europe: Game Boy – meet Tetris!

An even greater success during the history of games was Pac-Man, which was created 1980–1981 and inspired by a pizza with a slice missing. Namco designer Tohru Iwatani went out for the evening with some friends and their choice of food has had a major impact on the gaming industry.

Together, Tetris and Pac-Man (and others in the same bracket) conquered a new generation of game players and have shown the immortal nature of casual gaming. For Game Boy, Tetris was one of the key success factors, already showing how great devices thrive together with great applications.

During the next 17 years, new Game Boys have been introduced, but at a comparably slow rate. For example, it took 10 years for the first Game Boy Color to arrive in 1998.

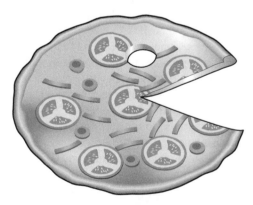

Figure 1.11 The recipe for success for mobile gaming?

In 2000 the Game Boy fans celebrated 100 million units having been sold, and the story goes on with Game Boy DS being introduced in 2005. Nintendo has time after time showed strength in making things simple but addictive and fun, things we have found are also key success factors for mobile phone games. Usually, people take out their phone for 5–15 min to kill some time while waiting for the bus or in a boring class. Tetris and Pac Man epitomize this gaming spirit.

The PC and console business has had a much quicker development in terms of features and richness. While console makers Sega and Nintendo dominated games earlier (with a little bit of Atari), Sony's release of Playstation in 1995 (US and Europe) changed the scene. Sega finally stopped making hardware and began making games for the other consoles. Playstation's appeal has been wide but, if one thing is to be singled out, it is the broad selection of games. The PC and console business became even more interconnected when Microsoft entered the console space with the Xbox, and this console was quick to enable network gaming, PC-like development and some top quality games. Xbox has therefore captured a significant market share and today (in 2005) the battle rages on between Playstation 2/3 (Sony), Xbox/Xbox360 (Microsoft) and Gamecube (Nintendo). One important development in this area is how the increased graphics and sound capabilities have rapidly increased the costs of development. Today it is not easy for a small, new entrant to break into this consolidated market and just a single failed game can lead to bankruptcy.

An important take-away from the gaming industries is how stable and predictable the platforms are. The latest Game Boy is backwards compatible with all the previous ones and therefore the addressable market is huge for new games. Console games show similar stability, with Playstation 2 (PS2) being interoperable with Playstation 1, and developers have been working with the same PS2 platform for several years. For these developers, the mobile phone market is scary, with hundreds of new devices every year with varying degrees of standard compliance and interoperability.

Is the mobile gaming industry going to be similar to the console business, where 10 % of the titles generate 90 % of the revenues?[11] We have already seen similarities in that the development cost increased rapidly to support 3D graphics and surround-sound capabilities in the devices (from €10 000 for the first games to over €100 000 for advanced games today). Or perhaps we will see more of the digital TV usage patterns, where Tetris, PacMan, Solitaire and other simple games are the most popular ones. This is not only due to the low capability of the set-top boxes but also to the way customers use it for casual gaming sessions, similar to Game Boy.

Actually, we see both of these tracks developing within the mobile gaming segment – advanced games with rich graphics (sometimes with multiplayer capabilities) and casual games to kill a few minutes of time. Already we have seen how the casual ones attract a large portion of the gamers, who are also prepared to pay for the games (see Table 1.2, showing the UK's top selling Java games in August 2004, with two casual games at the top and five more further down).

Some of the large console game and movie franchises, like Tiger Woods, Top Gun and Spiderman, also generated large revenues in the early days of more advanced mobile

---

[11] Source: Strategy Analytics report.

Table 1.2 UK top- 10 mobile Java-enabled games, August 2004.[12]

Rank	Title	Game maker
1	Pac Man	Namco
2	Tetris	Ifone
3	Racing Fever 2	Sumea
4	Mafia Wars	Sumea
5	Pub Pool	Iomo
6	Street Soccer	Sumea
7	Collapse	Jamdat
8	Steve Davis Snooker	Iomo
9	Space Invaders	Digital Bridges
10	3 in 1 Arcade puzzle pack	AMS

gaming. For that category cross-marketing with other media plays a big role: if your game has the same name as a movie with a US $30 million marketing budget it is clearly easier to get attention for the game. This ties into a concept that we have seen growing significantly lately – the way the success of mobile services depends the content's success in other media. The same has been true for other cross-media offerings (movies, console games, music, etc.) and perhaps indicates that the mobile channel is growing in maturity as a distribution channel. As an example, the game developer Synergenix in 2004 changed the name of its successful rally game to Colin McRae rally (after the famous rally driver) and saw the sales quickly increase 10-fold.

A crucial challenge in mobilizing a game that is connected to a known brand or movie is to ensure that the customer experiences are connected. The content has to be presented to the customer in a context that is relevant. For DVDs this is easier: the cover has pictures of the actors and photos from the movie. This means that the customer is presented with a context in which buying this DVD is tightly connected to the content of the DVD (i.e. the movie or TV show). In the early days of mobile services this has been harder. If the context in which the customer is presented with your game is only a text menu (three to five clicks away from the starting portal) with only the name of the game, it is harder to raise the same feelings and associated emotions to the movie that were obtained with the DVD. This is not only applicable for games connected to movies but also to cartoon figures and console/PC games. Examples of getting around this challenge include devices with richer graphics/screens as well as dynamic and rich portals, but also alternative distributions. Just as we see console games bundled with the top games, phones are sometimes bundled with games and ringtones from a movie (e.g. a Tomb Raider phone package) The KiBi cards are another way, where the physical look and feel of the card can easily be given the same feeling as the brand it is associated with.

---

[12] Source: ELSPA, based on 3, Vodafone, Orange and T-Mobile data. Used mainly for illustration as the chart varies depending on operator promotions.

While the rapid growth in the mobile gaming market is good news for the whole value chain, it also brings a risk. First of all, the market becomes crowded with many players. This is one of the reasons why mobile operators increasingly are using aggregators to filter and assemble gaming offerings. Second, there is a risk of mistaking this for one homogeneous market rather than several diverse segments. This is why several examples of simply taking a console game or digital TV game mobile have failed miserably. The sales figures in mobile gaming, however, are showing how those that execute well are reaping the benefits.

1.4.4 TV channels (media) and producers (content) enabling mobile TV

While there has been lots of hype around mobile TV, here we would like to look at it from the media and content perspectives, with mobile as one or more distribution channel. For content players, the game is about maximizing the value of the content available and specifically the content where the company in question has the full rights. As an example, a TV channel (media company) might have a number of series like *Seinfeld* and *Sex & the City*, but the associated rights are only to broadcast the shows on that TV channel. When it comes to other aspects, including mobile content, the production (content) company owns the rights. For media companies that create their own content, things are of course easier. This could include a sitcom or soap opera (although several of these are developed by global providers, like Endemol for Big Brother), or stand-up comedy.

TV channels have been pioneers in exploiting the mobile channel as means for increasing the interactivity with the viewers. Despite the availability of digital set-top boxes containing uplink channels – i.e. the possibility for viewers to send messages to the channel/cable company via the remote control – the uptake of such services has been slow. In contrast, those TV channels that have enabled interactivity via mobile phones (initially mainly via SMS services, like voting) have often been very successful.

One of the first conclusions one makes when mobilizing content for a TV channel is that the mobile most of the time is the secondary media channel, not the primary one. Do not try to compete with the TV! If you watch a football game on TV, the real-time results or video instant replays represent a complement (not a substitute) to the TV channel and should be designed to be so. In this particular example (a football game), we have found that the initial focus on attempting to offer near real-time videos of the scores was not always what customers were asking for. The target customers, those interested in the service, were either watching a TV or the game itself or unavailable to look at it. In all those scenarios, the preferred service was to receive the scores as simple updates (e.g. text) and then get a video highlights package afterwards. Alternatively, an interactive mobile TV solution could be designed, as described in Chapter 4, where the mobile is clearly positioned as the secondary media channel that adds the features it does best – mobility, interactivity and personalization.

Another key aspect of the mobile channel that appeals to TV is mobility. Here we will show two examples of how the media owner can use the mobile to increase the time during the day when the viewers can access their favorite content. An intuitive model is of course just to take the same content that is running on TV, the TV channels, and stream them directly onto the device, as was pioneered by operators in Japan but also in the USA.

Case study: taking TV mobile – Sprint MobiTV

Several mobile services have been slower in take-up than the always-optimistic telecom industry predicted. A pleasant exception occurred with the US operator Sprint's launch of MobiTV, as early as November 2003. MobiTV was launched on a 2.5G network in times when most operators where expecting video-like content only to be viable in the higher bit-rate 3G systems. 1xRTT is an evolutionary step between the cdmaOne (2G, similar to GSM) standard and the full 3G systems of that standard. Owing to the limited network speed, MobiTV offering started out with only one to two frames (changes of picture) per second (normal TV uses 25 or 30 frames/s). The solution saw much quicker take-up than expected (stated both by Sprint and the TV channels involved), even though no official subscriber figures were released initially.

MobiTV was shortly afterwards also launched by the competing operator Cingular,[14] using the same application provider, Idetic. Given that this solution required little change in the operator networks, it was easy to get it running to test the customer feedback.

This service appealed to the customers because it was very familiar. The segments reached were very attractive to the carriers: young adult and mobile lifestyle power customers.[15] The first group included those addicted to soap operas who did not want to miss the first 10 min when they were going home on the bus in the evening, while the power customers used it more during the day, getting updated with news.

In August 2004, Sprint then launched a follow-up service, Sprint TV, with higher quality (15 frames/s) and more content; see Table 1.3 for Sprint's initial Mobile TV offerings.

In terms of really utilizing the mobility, let us look at how the specifics of the mobiles can be leveraged for video and TV content. While the PC users who have broadband at home get higher speeds, the 3G networks offer an interesting capability with their videocalls: video with guaranteed speed/quality for the customer. While this was initially mostly aimed at videocalls from person to person, it can also be used for innovative video services, like the Big Brother service first launched by 3 in Italy.

Table 1.3 Sprint's initial mobile TV offerings.[16]

	MobiTV	Sprint TV
Quality	1–2 frames/s	15 frames/s
Content	NBC Mobile, ABC news, Fox sports, the Weather Channel	Movie trailers, basic channels and premium channels
Price	US $10/month	US $10/month for basic package + US $4–5/month for premium channel

---

[14] AT&T Wireless also launched MobiTV in 2004.
[15] Source: Factiva report, The Golden Age of Take-Out, 14 March 2005.
[16] *Wireless Week* 15 October 2005 and Ericsson's Traffic and Revenue Growth program.

Case study: Grande Fratello (Big Brother), 3 Italy

The Big Brother TV show has become a classic reality show, with the straightforward format of a number of people living isolated in a house under the surveillance of TV cameras running 24/7. While many countries have followed suit, the live camera 3G service was first implemented in Italy. It gave the viewers the opportunity to make a video call (shortnumber 4884) directly to the cameras of the house, see Figure 1.12. This is a different customer scenario from the TV, where someone else chooses what the viewers should see. A selling argument was that this service was uncensored, showing whatever happened in real time. The service generated 20 000 connections during the 3 months of the series, priced at €0.9 for a 5 min connection. This service's ease-of-use and intuitive connection to the TV counterpart gave it great appeal. It also was not trying to compete with the TV format, but rather complemented it by also letting customers check on the action when not on their sofa.

From the media content owner's perspective, the approach has usually been to first mobilize the existing available content (audio clips, video clips) and from there go into

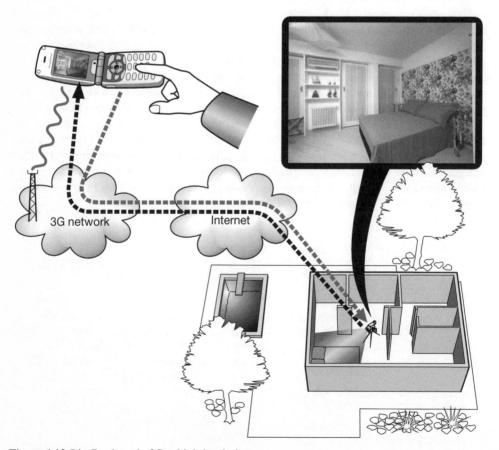

Figure 1.12 Big Brother via 3G – high-level view.

more interactivity and real-time aspects. However, it is key to evaluate these steps in terms of customer needs and business viability rather than technical advances. It is tempting to think that real-time streaming, combining voice and data are the last steps just because they are most advanced from a technical perspective. From a media and content company perspective, video content has much appeal because it is what customers and producers are familiar with (and lots of it is already being created). Therefore, some see video services earlier in their roadmap than MMS services, for example.

Key ideas that have been learned from mobilizing media from TV, movies and other video content include:

- *Learn by doing* – get something out there to generate feedback and experience;
- *Get small but positive results* from the first steps – avoid high-risk projects in the beginning in order to keep customers and stakeholders happy and calm;
- *Maximize the addressable market* – once moving into full-scale commercial service, choose devices and solutions that address a large audience that has an identified need and early-adopter/pioneer characteristics;
- *Integrate the mobile channel* – put mobile into the regular content creation process and always think about mobile rights when acquiring TV/video content (a live stream was previously considered under the umbrella of TV rights but not any more).

Going into the deeper sections of this book, we will then look more into the technical and commercial aspects of how to realize these solutions and requirements.

1.4.5   Gambling, betting and lotteries

The digital and connected setup of the Internet turned the whole gambling industry around. The regulatory environment is gradually changing, which means that individual countries are losing control over the revenues associated with state-operated gambling, betting and lotteries. With a customer betting on an Arsenal game using his French mobile while on vacation in Italy, which country should tax the transaction? The traditional rules of where to tax the transactions of the services become difficult in these scenarios. The situation is similar to the above-mentioned rights clearing for music content, like ring tones, and overthrows many of the established systems and value chains. This means that online gambling companies are rapidly moving their businesses to remote jurisdictions (like Gibraltar, Antigua and Malta) that provide favorable game license and tax conditions. From there the games can be offered to any player around the world. Individual countries are facing an uphill battle to defend their monopolies.

Furthermore, in the Gambelli case (Italy), a company outside Italy claimed to have the right to sell gambling offerings in Italy, referring to the EU open markets. The state countered and argued public health as a reason for not allowing the service. The EU court proved that it was not health but money that was the Italian motive, asking why there are such large marketing budgets for these services if the goal is to control them for public health reasons. However, we will continue to see these kinds of battles and some nations have chosen to prohibit the marketing of lotteries and betting that are not offered by the state (e.g. Sweden). In the end, this will probably lead to more possibilities for distribution of these services.

The global nature of the Internet has also given this business an extraordinary market share. Partypoker.com went from start to US $600 million in revenues (US $391 million in profits) in only three years, showing some of the potential. This company entered the London Stock Exchange 2005 at a valuation of US $8.48 billion, rivaling companies like British Airways at the time. During 2005 these online players accelerated their move into the mobile space and quickly spurred interest in the market. This industry is going mobile and it will generate money from this channel, which has the potential to become its biggest channel. Now the question is more about who will get a piece of that action. Will mobile operators be in this value chain? In this industry there are some characteristics that the telecom players need to adjust to in order to succeed.

First of all, these lotteries and betting services give a much larger part of the revenue back to the players (customers), in many cases well over 90 %.[17] This figure, the *Pay-out-rate*, is a key competition area – gamblers want to play where they have the greatest chance of winning. Now imagine that someone wants 30–50 % of the revenue for providing billing (a common revenue share demanded by a mobile operator); this is clearly not a feasible model. All online providers of these services have accounts setup for their players where they can transfer money via credit cards or from their bank accounts. Secondly, the online gambling industry has a tradition of working with affiliate models, where the one that links to a gambling provider can get a kick-back from the revenue generated by that player (a percentage of future usage).

Consequently, mobile operators that want to make money from gambling services need to either provide billing capabilities that fit this business model and/or leverage its marketing and customer reach to collaborate as affiliates. For the billing aspects, the challenge is two-fold:

- offer simple billing for transactions at competitive revenue sharing rates; for this kind of transactions, the credit card companies typically charge 3–6 % commission;
- provide possibilities for the customers to withdraw money from their accounts when they have won (reverse billing); alternatively, position the offering of alternative prices.

The other alternative, being a marketing affiliate, means that an operator links the customer to the gambling company's services, for example via its portal.

Case study: betting and gambling on the move – Mansion
Mansion is a large online gambling company with many plans for the mobile channel as part of its global ambitions. They quickly saw that, in order to tap into the 4.5 billion Asian population, for example, the mobile phone has the potentially highest reach, by far surpassing the PC and fixed Internet.

In 2005 Mansion launched its mobile service offering, based on the same back-end applications and gambling services as found online, like casino games, probability games and sports betting. From the beginning, the ambition has been to provide a superior and seamless customer experience between the different channels. Customers normally go to the web page (www.mansion.com), select the game they like and input their phone

---

[17] Source: gambling company interviews.

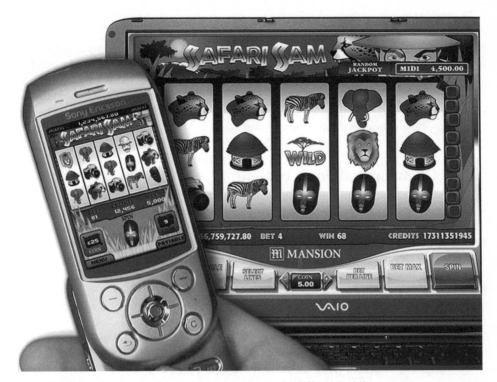

Figure 1.13 Mansion slot machine game (Safari Sam) for PC and mobile.

number. An SMS is sent with an installation link, which the customer clicks on. This triggers an advanced system to find the right software version for that device and send it to the customer. The customer, however, does not have to do anything more than watch the game get installed and start playing. Customers can start to play for free (e.g. slot machines, black jack) and later start getting money using an existing Mansion account or create a new one.

As seen in Figure 1.13, the web and mobile clients are very similar in appearance and user interface. In fact, games can be started on the PC at home with the web client and then the customer can continue the same session on the mobile when leaving the house. This is truly fixed–mobile convergence and makes it very natural for customers to extend their interest in the services to the mobile.

This case also illustrates how the convergence of industries induces new requirements for business models. It shows how new technologies give possibilities to launch services with or without operator involvement. In this case, clearly operators still have good opportunities to work with gambling companies in the area of casual usage. This is due to the fact that the existing gambling billing mechanisms with dedicated accounts attract mostly current online customers and die-hard players. The mobile operators have the opportunity to expand to casual users who would like to buy a lotto ticket or bet €2 on their favorite football team. The mobile phone bill is still a superior way for the customer to pay for small amounts, without having to open a dedicated gambling account.

The area of ethics connected to gambling has always been one of the first to arise in the telecom industry – how could this damage my brand and how do I prevent customer addiction? Gambling companies claim that this is already one of the most audited and supervised industries. For example, you need both a passport and a signature to withdraw money from a gambling account. The key for these companies is to be able to check the age of the players – if they are of legal age they can play. The comparison is made with a normal department store, where betting and lotteries are sold in many countries without any damage to the brand or reputation of that store.

1.4.6   Glamour and adult content

While gambling can appear to be a tricky area for operators to work with, it gets even more challenging with adult content and its lighter version, often called 'glamour'. This industry has always been the leader in adopting and making money out of new technologies, as shown with video recorders in the home and the Internet. Similar to gambling, several players here are accustomed to working with dedicated payment accounts on the Internet that can also be utilized for the mobile channel.

In this area you will find the established players, like Playboy, but also newcomers that are profiling themselves via the mobile channel. The two approaches are often complementary (which is not unique for this range of content), with the established players having enormous quantities of content but little that is adapted for mobile, while mobile-focused entrants build their content from scratch with mobile in mind.

Case study: Great Media taking new models to the mobile
Great Media, a Swedish content provider, has been leading the development of 'light' adult content (but also nonadult), including swim-suit models and topless men and women. An interesting differentiator has been the associated web sites in Sweden for talent search (www.girls.se, www.men.se) and opportunities to be discovered. Together with the operator 3 they early on started to tailor-make videos for the mobile channel and today over 100 operators and media players use their content. The take-up rates have been impressive, and Great Media has grown 10 % on a *monthly basis* for over 2 years. Having a strong policy not to produce hard-core material, Great Media has been an easy entry point for operators that feel hesitant about this content area but want to try their way forward. Another key success factor has been the effort put into making the content themselves in order to both control all the digital rights and to create the best possible quality for the mobile channel.

Actually, the popularity of the models that Great Media offer has reached considerable heights, e.g. in the UK two 'mobile models' have appeared on a model chart where normally only celebrities and models from TV and the printed press are seen. This is another example of the cross-media channel effects that we are starting to see as the mobile channel grows in importance. Already today, many models see the mobile channel as an important part of building their brand and career.

How should operators and content aggregators position themselves in this aspect? Again, using the department store analogy, it is quite normal in many countries to see

Sports Illustrated Swimsuit edition as well as adult magazines offered in regular shops. Several investment firms (fund managers and venture capital) have declared in their investment policy that they distance themselves from the production of adult content but not the distribution. Cable TV providers are another example of distribution channels that offer adult content (of much more explicit nature than the mobile examples above) but do not produce it. As with cable TV, it is recommended to introduce mechanisms for parents to block certain contents for their children.

When looking at the key requirements from this industry on the mobile channel, they have most of it in common with any other content and media industry segment, i.e. good payment and accounting systems supporting fair business models, digital rights management and a significantly improved situation on the number of formats to support. Great Media, in the example above, has to support over 350 different image formats and 15 video formats to fit with the requirements of devices and service providers (operators and media companies). The devices are indeed part of this, with many screen sizes, color palettes, etc., but perhaps even more time-consuming is that each media company, operator or aggregator sends them a list of 30+ formats to support – each one with different requirements for the same or similar devices. As a solution, Great Media has built a conversion tool that does this automatically, but still it consumes much administration.

Summarizing, there are many common factors between glamour and adult content and other kinds of media, but perhaps the customer demand is the highest of all segments. In the early WAP days a German mobile site of adult content generated as much traffic as all users in Turkey combined, and for one big European operator, Playboy has been generating two-thirds of all video revenues.

1.5   MAKING A BUSINESS OUT OF IT ALL

The starting point for looking at these value chains is that all three to five players want 60 % each, which makes for an equation that not even the most skilled math professor can satisfy. We have seen how content companies come from a world of large upfront payments for usage rights and operators have a unique position of controlling the billing relationship with mobile customers. The first thing to realize is that there will probably not be one single business model but that it will be dependent on the service and the players involved. Secondly, this is an area where no one is standing on firm ground and the conditions will change greatly in the coming years.

1.5.1   Operators and their business models

The mobile operator has from the start been at the center of the value chain, controlling access to the customers. During the early hype-years of 1999–2001, we saw great diversity in the business models adopted. The concept of a 'bit-pipe', the role of only adding value by transporting the information to the customers, was the worst-case scenario that everyone feared. Consequently, different ways to climb the value chain and do more were explored. Several of these attempts to reach into new areas failed, not only due to bad execution but also because of the immaturity of the whole industry, devices and the offerings to the customers. As the industry has advanced, so has the notion of the ideal business model.

First of all, let us look at the different revenues from mobile services in order to establish the basic concepts. The total cost of a download (e.g. a game or a ringtone) has traditionally comprised both a traffic fee (paid per kilobyte of data or per SMS) and a content fee (e.g. €1 for a logo/wallpaper). For big files, like video clips, the traffic fee can be a significant part. One of the key success factors for the mobile Internet in Japan has been the strong push to achieve a working business model. As mentioned above, NTT DoCoMo takes only 9% of the content fee from its official partners and the majority of its revenue comes from the traffic charges (see Figure 1.14). Despite this often being claimed as a success story, most of the content providers in Japan are struggling with their profitability and are campaigning to also share some of the operators' traffic revenues. Traffic from unofficial sites in early 2001 had already exceeded traffic from official sites,[18] which shows that people want more than just the services provided by the operator.

The European (and US) operators have from the beginning takes a large portion of the content revenues in return for charging capabilities and placement on their portal. Here the development has taken a different direction than in Japan, and less of the focus has been on the traffic revenues and more on the content charges. As an example, Vodafone in late 2004 introduced a scheme where navigating within Vodafone Live would be free

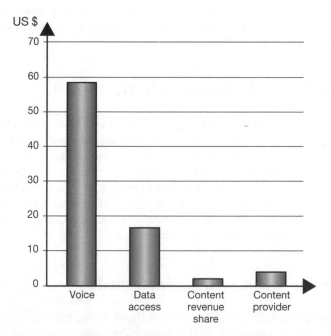

Figure 1.14 Most of the revenues in Japan come from traffic charges[19].

---

[18] February 2001 official traffic was 45% of traffic, and unofficial was 55%. Source: J. Funk, *The Mobile Internet: How Japan Dialled up and the west Disconnected*. John Wiley & Sons: Chichester, 2002.
[19] Source: analysys. Note: in December 2003, 71700 sites were available on the i-mode 'web'.

(no traffic charge) and customers would only be charged for the premium content. The contradiction here is that the European system of asking customers to pay for the content they desire sounds more intuitive, while the Japanese operators are clearly proving that they can make money on the traffic generated by the services.

Another part of the mobile operator business model is connected to the brand and the desire to differentiate from the competition. When saturation appears in a market (which is the status of many Western markets), the operators compete partly on the price of the basic voice service offering and partly on the perceived brand image and the associated services. Owing to voice still being the main source of revenues, anything that can help an operator attract or retain customers who make lots of phone calls forms a central part of the strategy. In fact, voice in some markets has such good margins that operators feel reluctant to launch other services, claiming that their customers spend only €25 per month and want maximum margin (i.e. voice margins) from that revenue. We have, on the contrary, seen several examples that show that customer spending is rather flexible, as will be shown in Chapter 9.

In the days where only voice (and to some extent SMS) existed as service, there was little need for partnering for mobile operators. With the mobile channel becoming increasingly interesting to other industries, the diversification of the service offerings is increasing (music, mobile TV, etc.) and consequently also the need to have expertise in those service areas. By the nature of many industries we have looked at in this chapter, one company cannot be the expert in all of it. Even the big record labels are humble before the challenge of launching a new artist and frequently use outside help (consultants) to maximize the chance of success.

In this environment, new operator business models have emerged, and we have divided them into three models:

- *service provider* – the operator offers the service under its own brand (e.g. Sprint voice-mail), sometimes by purchasing content or applications as part of it;
- *channel provider* (sometimes called Smart Pipe) – where the media or content company addresses the customers directly but additionally the operator offers channel capabilities, like billing, positioning and identity management, and charges for these capabilities;
- *bit-pipe* – where the operator revenue only comes from the traffic, providing network access, and the customer gets the service directly from the media and content companies, e.g. news from CNN or the gambling example above.

Rather than seeing one of these models as being the one and only, we are likely to see different models for different services. An operator probably wants some core services branded with their name (voicemail, picture messaging, etc.), while we will increasingly see other models for media and content.

It seems that all a service provider (operator or media company) has to do is to decide on the business model(s) and start with the services. Here comes the catch: these are often difficulties and inefficiencies when trying to migrate to new business models using the same roles and processes. Therefore, a full commitment to a new business model needs to be joined by a full commitment for implementing it in the organization. New demands include cross-functional processes to launch services (sales, marketing, technology, IT), service fulfillment and billing/accounting in collaboration with the new value chain, to name a few. It should be stressed, however, that the bit-pipe model is not a bad business model and is much less demanding than the others, so some operators might choose to stay in that area.

These challenges contribute to the trend for mobile operators today, either in an evolutionary or revolutionary way, to enhance their service environments to cope with these new requirements. Similarly the other players in the value chain, like content and media companies, are evolving their capabilities in terms of content-delivery infrastructure and application servers. A term sometimes used is *the service layer*, referring to the sum of the business and technical components that are needed to realize mobile media and applications (including devices). Any specific implementation of such an environment is commonly called a *service environment, Service network* or *service delivery platform* (SDP). In Chapter 7 we will go deeper into the aspects of building and optimizing such service environments, coupled to the business models.

1.5.2   Aggregators – linking content and distribution

In the model outlined above, where operators have been the way to reach the customers and get paid for the content, it becomes difficult for all media, content and niche developers to build relations to all those operators. Additionally, it has become increasingly demanding for operators to filter all applications that come their way and have competence in all the service areas that they need to provide. Because of this, the role of content aggregators has emerged in order to collect strong portfolios of offerings from developers and media companies and distribute them to operators around the world.

The role of an aggregator is therefore one or more of the following:

- investment in services connected to big brands (e.g. Spiderman, Fox);
- handling and hosting the technical complexity;
- own development of part of the content;
- sourcing of development from more specialized studios;
- distribution to those that have customer relationships (usually operators);
- expertise in different service areas, like games, music, etc.

One part of the role of a mobile aggregator is similar to the publishers existing in the gaming industry, which build up production and distribution on a large scale and thereby are able to lower the risk per offering. Similarly, it is very tricky for a small development company to deal with the big content players, like movie studios, in acquiring the rights to offerings around such franchises as Star Wars or Spiderman. The license fees – both their size and the fact that they are asked upfront rather than by revenue sharing – requested by the big content brands were a source of controversy in the early days of the industry. In fact, the upfront fee for a creating a game based upon a blockbuster Hollywood movie early on could reach over US $1 million. Today, the levels have come down somewhat, but still present a major entry barrier for new content aggregators.

With the limitations of the early versions of WAP and associated devices, many media companies have been frustrated by bad customer experiences and have therefore been hesitant to mobilize their media. A major role of the content aggregators, like ringtone providers, has been to solve these technical issues and convert raw content into appealing mobile services. The challenge will be to provide additional value on top of this the day that the industry moves beyond these initial hiccups. Sometimes making it easy for the customer requires much complexity behind the scene, e.g. solutions for ringbacktones that combine intelligent networking integration with content management (see Chapter 8 for a case study on this topic).

Will aggregators be less important in the future? Will operators and media/content players do business directly or will perhaps the media companies take over the aggregator role? This remains to be seen, but there are clearly many areas of potential value added between those that focus on the content and those that focus on the distribution. One example is helping to understand customer behavior by monitoring usage and skilled processing of the feedback (something that media companies are paying big money for in the TV and radio area, to draw a parallel). As the value chain grows in maturity, we are likely to see companies focusing more and more on their core business. Already today, operators are moving from technical management of these solutions to focusing on the marketing and product management aspects. Probably, hosting and content management is not going to be core business for operators, media or content companies.

1.5.3   The developer in the middle

If aggregators find life in the mobile service value chain difficult, it is often even tougher for small developers. The developers are generally more focused on creating the technologies and solutions needed, specializing in vertical areas like music or games (although it is common that a company is both an aggregator and a developer, like Mforma). While there exist large developers, like Ericsson and Nokia, we will here take a closer look at the small and medium-sized developers because of their role as innovators. Without companies with a passion for innovation and new technologies and solutions, the evolution of this industry will be much slower. The difficulties for developers come from the same causes as those described for aggregators, in that operators on one side and content players on the other have strong positions with their control of distribution and content. However, these are dependent on good games, good technologies or innovations like recognizing radio songs via the phone microphone in order to evolve their offerings.

Case study: innovative developer seeks funding
To protect the innocent, we will call this developer SuperInnovation when describing its struggle in today's marketplace. SuperInnovation had start-up (seed) funding from the owners and some friends to develop a ground-breaking application in the mobile TV area. In order to obtain further funding they would need to show that both the technology and the business model would work and be scalable. Having a first version ready, these entrepreneurs approached telecom vendors and operators to jointly perform trials with friendly customers. Finally, an agreement was reached with Operator X to go to market, but only after tough negotiations that left SuperInnovation only 15 % of the revenues after the media, content and operator companies had taken theirs. Additionally, SuperInnovation had to pay all the costs for adapting and integrating the solution for the specific solution and the revenue share was to be paid out 120 days after usage. The founders

thought that the conditions were bad but knew that they had to get something to the market to prove their value.

The trial went rather well and SuperInnovation looked like a promising company and went to venture capital (VC) players to look for further funding in order to expand. The VCs, however, did not at all see the business viability in SuperInnovation as a company due to the terrible revenue sharing they had obtained in the trial. Thus, SuperInnovation kept struggling and finally obtained some more funding with very bad terms, but their entrepreneurial heart and enthusiasm were seriously hurt.

So what can developers do to tackle this difficult situation? Throughout the case studies in this book we will see some good and bad examples of strategies and execution that can serve as inspiration. Let us mention, however, some key aspects:

- add value all the way, not just through technology but also by helping partners/customer understand the impact on marketing and customer needs;
- make things simple, get something out on the market and start generating feedback;
- align with the interests of partners/customers, for example by helping a vendor boost IMS sales by providing attractive applications that are not competing with the platforms.

We must, however, not forget the needs of the developers and contribute to creating a fair value chain where innovation pays off (at least if the innovation adds value to the customers).

1.6   SUMMARY

These are indeed interesting times, with several powerful industries coming together to form a marketplace. Since all of these players come from different industries, they bring different characteristics to the mobile channel. These range from requirements to control content presentation and clearing rights to having an easy way to create applications for multiple devices. Finally, the new business models that are emerging require a lot from all players in the value chain and the race has just begun.

2

Understanding the Customer

2.1 WHY UNDERSTAND THE CUSTOMER?

A simple question, and one that many would think was trivial: after all, if we have a product to sell surely we would like to know something about the people who might buy it? However, with a few honorable exceptions, the mobile communications industry has been very poor at understanding its customers. This is perhaps forgivable in a business whose performance has regularly outstripped even the most optimistic predictions. The biggest challenge for mobile providers has often been keeping up with demand, and when your customers are banging on your door demanding to buy your services you do not stop to ask why!

As technology has advanced, though, mobile networks have acquired high-speed data capabilities and mobile handsets have become more than just phones – they are now hand-held computing devices equipped with a keyboard and display, multimedia capabilities, unique features such as the ability to identify a person's physical location, and an interactive, wireless interface to a global data network. As a result the media industry now sees the mobile phone as an exciting new route to 'its' customers and is busy developing mobile versions of music, movies, TV programs, news, and so on.

All this opens up infinite opportunities for new mobile applications but at the same time it makes the market for mobile services far more complex. As long as voice telephony was the primary service offered, the only thing mobile providers needed to know about their customers was whether they could pay their bills. Today a mobile operator may be offering an entire suite of content and applications, from messaging to multimedia entertainment and multiplayer games. It is easy to see that different mobile applications will appeal to different people; it is critical then to match each application to the customers who will be most likely to use it, and this is why we need to understand our customers.

Chapter 1 introduced us to the history of mobile telecoms. Let us pick up some of the key milestones in this history and see what we can learn about how, and why, customers use mobile applications.

Mobile Media and Applications – From Concept to Cash: Successful Service Creation and Launch   Christoffer Andersson,
Daniel Freeman, Ian James, Andy Johnston, Staffan Ljung
© 2006 John Wiley & Sons, Ltd

2.2   MOBILE APPLICATIONS – THE CUSTOMER'S VIEWPOINT

2.2.1   Voice – cutting the cord

The first mobile application was voice telephony, in other words the ability for people to talk to each other. Voice is the application that has grown the mobile business to what it is today – a US $500 billion industry with around 2 billion customers. Today, voice continues to generate the most traffic and income for mobile carriers, and most likely will always represent the core of any mobile operator's business.

It would be a mistake, however, to assume that the success of voice telephony was inevitable, that today's insatiable demand for mobile phones was ever present, and that people were simply waiting for engineers to invent the technology that would burst the floodgates to a new mobile lifestyle.

It is easy to forget that, as we saw in Chapter 1, it took up to 13 years for mobile phones to reach even 25% penetration. Of course the first mobile handsets were bulky and expensive and were only bought by people with a real need for mobility, such as traveling business people and perhaps certain elements of the criminal fraternity! Growth accelerated as the size and price of mobile handsets decreased with the introduction of digital services, but mass-market penetration only occurred when operators introduced prepaid services and fierce competition drove down the price of calls.

So let us not take success for granted. The mobile industry has been created over a long time and with much hard work. It provides a great place to start for providers of mobile applications but we too must work hard to ensure that our customers see the benefits of our lovingly created products.

2.2.1.1   What we can learn: customers are resistant to change

The time it took to establish mass-market mobile telephony was partly due to the early cost and technical limitations, but it also took us a long time to recognize and embrace the benefits of the mobile phone. There was a fear that our peace and privacy would be invaded by uninvited calls. Today, however, as established mobile customers, we feel naked and vulnerable without our mobiles. Our attitude and behavior have changed entirely, and this perhaps explains the time it took. Attitudes and behaviors take a long time to change. We are creatures of habit and we frequently resist change unless there is a clear need or advantage to doing so.

2.2.2   SMS – a 'killer' application

Hype and hyperbole are standard marketing tools of the communications industry, but it is difficult to exaggerate the success and influence of SMS. Almost an afterthought in the original GSM specifications, the 160-character Short Message Service has today become an integral part of millions of people's lives.

SMS use is highest in Europe and Asia but is following a similar growth pattern, albeit later, in the rest of the world. In the UK alone, during 2004, 26 billion SMS messages were sent,[1] but even this seems modest when compared with the Philippines – the SMS capital of the world – where 67 trillion SMS messages were sent during 2004, equivalent to 169 messages per month for each of its 33 million mobile users.[2] Today as much as 20 % of a mobile operator's income comes from SMS messaging.

The ubiquity of SMS means that its applications have grown beyond basic person-to-person messaging: people use it to interact with TV and radio programs, book cinema tickets, pay for car parking, enter competitions and receive news and sport alerts. SMS use is so widespread in the UK that in 2003 citizens were given the option to vote in local elections using SMS. Following the Asian tsunami disaster of Christmas 2004, mobile subscribers in many countries were able to make donations to aid funds simply by sending an SMS to a short-code number.

2.2.2.1   What we can learn: young customers were the first adopters of mobile data

Young people were largely responsible for the early growth in mobile data services, with SMS leading the way. The emergence of the SMS generation gives us some insight into youth culture. The very difficulty of typing SMS messages meant it was unlikely to be quickly adopted by adults. SMS thus satisfied the need for young people to differentiate themselves from their parents and adult society, as well as providing a cheap, effective and – with a new SMS vocabulary – somewhat subversive way of maintaining group communications.

Unfortunately, no one in the mobile industry stopped long to consider why SMS had been so successful. In fact it came as a total surprise to many; business users had always been assumed to be the natural first adopters, and manufacturers were busy developing cellular data cards for laptop computers, packing infrared modems into mobile phones and promoting the business benefits of mobile data.

This was understandable; after all, it was business users that had driven the uptake of mobile telephony in the first place. However, in the early days these were the only people who could afford to become mobile; young people were excluded by the cost of handsets and the absence of prepaid. The removal of these barriers revealed young people as the true early adopters.

2.2.3   Ringtones and logos – the mobile phone as fashion

Quick to react to the growth in young mobile users Nokia started producing low-cost mobile phones specifically aimed at the youth market. While other mobile phone manufacturers continued to produce rectangular, functional (and mostly black) handsets, Nokia introduced stylish, curvaceous phones in a range of colors, clearly aimed at fashion-conscious teenagers. Some handset models even had interchangeable fascias, which allowed people to further personalize their phones.

---

[1] BBC News web site, 21 January 2005.
[2] EMC, World Cellular Data Metrics, March 2005.

Nokia then took the idea of phone personalization one step further. In 1997 the company announced a development of SMS called 'Smart Messaging', which allowed ringtones and screen logos to be sent to and between mobile phones using a standard text message.

Ringtones and logos enhanced the mobile phone as an individual symbol of fashion and status; it gave young people a new way to differentiate themselves from each other and the mass market, which strengthened peer-group identity. Moreover, the ability to exchange ringtones and logos enhanced the community aspects of SMS among teenagers and was instrumental in establishing the mobile phone as a youth icon, symbolizing freedom and individuality.

Around the same time Nokia established a web portal, Club Nokia, which contained content and services that its customers could download to their phones. Some mobile operators saw this as direct competition for control over what they regarded as *their* customers, but it spawned an entirely new industry for the provision of mobile content.

2.2.3.1 *What we can learn: understanding customers works!*

The popularity of ringtones and logos demonstrates the need for young people to express themselves, but the key thing to learn here is how effective customer understanding can be. Nokia clearly understood its customers. Rather than promote 'technology' to a largely disinterested public, it first spotted a new customer segment – young people; it identified the characteristics that defined the segment – fashion, music, the quest for independence; and it then developed the technology to satisfy those requirements. The results speak for themselves. Even today the company still retains around a 30 % market share of the global handset market, and ringtones is a multibillion-dollar business.

Nokia's introduction of Smart Messaging caused consternation in the mobile industry because it was a proprietary solution, i.e. it was not part of the GSM standardization. Nokia hoped other handset manufacturers would license Smart Messaging but most hung on for an industry-wide standard, ultimately published in 2001 as EMS (Enhanced Messaging Standard). By this time, however, Nokia was entrenched throughout the world as the mobile phone brand of choice amongst a generation of young people.

There is a down side to this. Proprietary solutions are always the fastest way of getting a new technology to market, mainly because, as a developer, you do not have to satisfy anyone else's opinions or needs. In retrospect, Smart Messaging kick-started the content industry 3–5 years before it would have happened with an industry-wide standard, but the result has been a fragmentation of ringtone and logo formats. For example, there are least four monophonic ringtone formats for different makes of handsets, and people with older Nokia phones are unable to exchange logos and ringtones with people with phones from different manufacturers.

Likewise there is today no overall consensus on the formats for wallpaper images, real music tones, games or videos. The consequence of this is that content providers have to produce their content in numerous different formats – increasing the cost considerably. Provision of content is also made more complex: when a user requests a content download, the content distributor must somehow detect the user's handset make and model in order to select and deliver the right content format.

It would be unfair to blame Nokia entirely for this state of affairs, but it does illustrate the challenges of multiple standards. As the old joke goes, the beauty of standards is that there are so many to choose from!

2.2.4 The mobile Internet – two approaches, two outcomes

By the late 1990s the number of Internet users and the number of mobile subscribers globally were each rising exponentially. Many in the mobile industry inferred from this, somewhat tenuously, that the combination of Internet and mobility would therefore make the perfect marriage.

It was already possible to connect a laptop PC to the Internet using a mobile phone as a modem. This targeted the narrow but potentially lucrative segment of traveling business people, but because the mobile connection was slower and more expensive than a conventional fixed telephone line, it failed to attract many customers. What was needed was a service, suitable for the mass market, that brought the Internet *directly* to the mobile phone.

Let us look at two examples of mobile Internet services with similar aims but very different outcomes.

2.2.4.1 i-Mode

NTT DoCoMo's i-mode permitted its customers to send and receive email and to browse a range of information, fun, entertainment and m-commerce services using the mobile phone. In its first year i-mode accumulated 4 million subscribers; after 2 years this had risen to 20 million (Figure 2.1). At the end of 2004, 43 million people in Japan subscribed

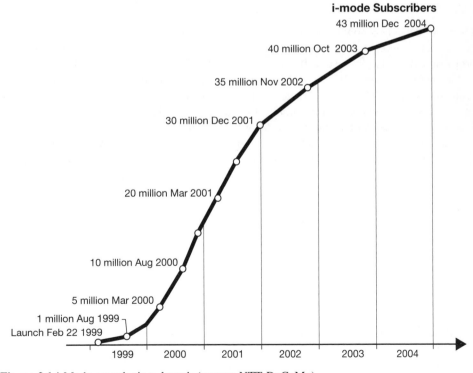

Figure 2.1 i-Mode growth since launch (source: NTT DoCoMo).

to i-mode, representing approximately 87 % of NTT DoCoMo's total customers and 43 % of the total population, and i-mode services were generating an average monthly revenue of over US$18 per user.[3]

2.2.4.2 WAP

WAP is a 'language' that allows mobile phones to connect directly to the Internet over the mobile network. Unlike i-mode, which only functions with NTT DoCoMo's handsets and network, WAP was standardized for use on any mobile network. It had the support of almost the entire mobile industry, thus ensuring compatibility between all WAP-enabled phones.

The first commercial WAP services were launched in Europe in 2000. Billed as 'the Internet on your mobile phone', operators spent lavishly on promoting WAP as the next chapter in the mobile revolution, only to be met with public indifference and media derision. The public, accustomed to accessing the Internet on high-performance PCs, was unimpressed with the predominantly text-based service, displayed on the small, black and white screens of mobile phones.

WAP never produced the mobile Internet the industry envisaged, but it has survived and is the basis for practically all mobile content downloads today. Indeed, during 2004 there were almost 15 billion WAP page hits in the UK – a sign that WAP has finally found significant market customer acceptance.

2.2.4.3 WAP vs i-Mode

The WAP débacle is now legendary and will probably become a textbook standard on how not to market mobile services. There is nothing wrong with WAP *per se* – as a wireless protocol it does its job very well – it is just that WAP should never have been marketed to the public as, well, 'WAP'. Imagine if the Internet were promoted by its underlying protocol TCP/IP WAP was marketed either as a technology, which few understood or cared about, or oversold as the 'mobile Internet', which clearly it was not. To a market already accustomed to the Internet, WAP was pathetic in comparison.

Other factors also worked against it: before users could access WAP services they were required to enter a series of obscure configuration parameters into their phones; connections were unreliable; response rates were slow; and with a WAP data connection charged at the same rate as a mobile voice call, the service was expensive for what it offered.

Content providers could share in WAP revenues, but mobile operators weighted revenue-share models so heavily in their own favor that there was little incentive for content providers to produce WAP content. As if this was not enough, many operators operated a walled-garden approach, which prevented users from reaching WAP sites outside the operator's own portal.

The experience of i-mode differs from WAP in almost every way. In contrast to Europe, Internet penetration in Japan at the time of i-mode's launch was relatively low; i-mode was therefore the first time people in Japan had experienced Internet-like services, which

---

[3] Source: NTT DoCoMo.

is one key factor in its success. However, equally important were the steps NTT DoCoMo took to refine every aspect of the i-mode experience:

- specially designed i-mode phones were commissioned, many with color screens;
- phones were preconfigured with all relevant parameters and each had a button that, when pressed, connected the user directly to i-mode; this made it extremely easy for users to access i-mode services;
- services were cheap – a typical service cost between ¥100 and ¥300 per month (approximately US$1–3), which was affordable for most people in Japan;
- i-mode was an open portal – the i-mode menu listed official content providers but users were free to browse to any i-mode-compatible site;
- CHTML – a version of the Internet mark-up language HTML – was developed that made it easy for content developers to adapt or create i-mode content; today there are over 4000 official i-mode sites and more than 83 000 independent sites;[4]
- a generous revenue-share model gave 91 % of content subscription fees to official content providers, thus providing a strong stimulus to the content industry; DoCoMo bills customers directly, thus relieving content providers of the task;
- DoCoMo marketed i-mode not as a technology but simply as cool things to do with your phone;
- last, but not least, i-mode was conceived by a nontechnical person. Mari Matsunaga was headhunted by NTT DoCoMo in 1997; an editor by training and a self-confessed technology illiterate, she envisaged a service that 'even she could use'.[5]

2.2.4.4 *What we can learn: customers like things to be easy*

Perhaps the biggest inhibitor for the adoption of mobile applications is the fact that too often they are too complicated for the average user. Even the handset itself can act as a barrier to use – for instance, customers should not be required to configure their own phones. Voice and SMS have reached the mass market largely because all a customer needs to do to make calls and send messages is insert the SIM card and switch the phone on. Conversely, the same customers will most likely not be able to use any online data services because the parameters must first be loaded to the phone. Most people have no idea how to do this or inclination to learn, and are thus prevented from discovering and experimenting with new services because they simply will not work.

In addition, many mobile services are still far too difficult to use for the average user. Assembling an MMS, navigating WAP sites, even SMS, all involve numerous key clicks and assume a fair level of computer literacy, which most people lack.

Simplicity is not easy to achieve; even one barrier is enough to prevent a service being used. i-Mode can be regarded as best practice in this regard. Conceived by a woman who found technology intimidating, she permeated the service with ease-of-use features, with outstanding results. It is also interesting to note that, whereas the mobile industry even today remains fixated with faster and faster bandwidths, i-mode achieved mass-market success with a snail-like data rate of 9.6 kbit/s.

---

[4] Source: NTT DoCoMo.
[5] 'The Birth of i-mode. An Analogue Account of the Mobile Internet', M. Matsunga, Chuang Yi Publishing Pte. Ltd. 2000.

2.2.5   Mobile e-mail – something for the business user

Mobile e-mail has become well established in recent years, largely due to something called BlackBerry®.

BlackBerry, from Canadian company Research In Motion® (RIM), is a wireless communication solution based around a family of devices (Figure 2.2). Launched in January 1999, BlackBerry initially became established among business users in North America, who used it to access their corporate e-mail wirelessly. BlackBerry has now been launched by mobile operators in Europe, Asia Pacific, Africa and Latin America.

The BlackBerry solution includes the devices as well as server software, hosted behind the corporate firewall, which interfaces to a company's corporate e-mail. Each device has a graphic display and a 'qwerty' keyboard and connects to a wireless packet data network. Emails are pushed to the handset as they arrive, using GPRS (general packet radio services) or other 2.5G technology, thus eliminating the need for users to connect and download e-mails. Data is encrypted to ensure end-to-end security.

The popularity of BlackBerry is due to a number of factors:

- The first BlackBerry device – the RIM950 – merged the two most common means of nonvoice communication in North America at the time: the pager and e-mail. A single hand-held device offered a significant advancement on the pager, as well as ubiquitous, two-way access to e-mail.
- Access to e-mail is simple for users and the solution is secure and simple to implement into a company's IT infrastructure.
- RIM established a large network of resellers, partnering with operators, Internet service providers and system integrators, who share the subsequent revenues.
- RIM work with potential competitors such as Intel, Microsoft and IBM to convert them into allies by incorporating BlackBerry technology into their own solutions. (© 2005 The RIM and BlackBerry family of related marks, images and symbols are the exclusive properties of and trademarks of Research In Motion – used by permission.)

RIM 950

BlackBerry 7290   BlackBerry 7100t

Figure 2.2 Blackberry devices.

2.2.5.1   *What we can learn: with business applications there are several customers we must understand*

Applications for business users have not been as prominent as for consumers. One reason for this is a business application has to meet the demands of several people:

- the individual business user, for whom the application must provide a clear benefit and be straightforward to use;
- the CEO, who needs to understand how the application will increase the company's productivity, reduce its costs and/or increase its efficiency;
- the IT manager, who is usually the person tasked with implementing and managing the application, and for whom security and reliability is vital; ideally the application should also be straightforward to install and require little ongoing maintenance.
- the financial director, who will wish to see a rapid return on the company's investment in the solution.

The BlackBerry solution clearly manages to satisfy all of these stakeholders.

Another thing to learn is the benefit of building on established customer behavior. The first BlackBerry device resembled a pager, which no doubt contributed to its quick acceptance by North American users. Also, people are accustomed to their e-mails arriving on their PCs without intervention, so the push-e-mail function of the BlackBerry (and other recently launched mobile e-mail solutions) again mimics this behavior.

2.2.6   MMS – a picture is worth a thousand words, or is it?

MMS, in a nutshell, allows people to send pictures, text, audio and video between mobile phones, similar in principle to SMS, although using different underlying technology. It can also be used as a download bearer, through which content can be requested by the user, or pushed by a content provider.

The first launches of MMS took place in 2002. After the WAP experience the mobile industry had been cautious about over-hyping new mobile services. However, all in the industry were sure MMS would finally be the breakthrough for mobile data. It seemed an obvious winner: MMS built on the success of SMS; sending pictures was a concept anyone could easily understand; pictures added an obvious benefit over text, for which a premium could be charged; and recently launched were the first color-screen GSM phones with built-in cameras, which would stimulate market adoption. Everything looked set for MMS to fly.

Three years later MMS has not been a complete flop but its growth has failed to match initial expectations. Even in countries with high messaging use, the number of MMS messages sent by individuals is still dwarfed by the volume of SMS messages. For example, mobile users in Denmark are avid text messages, but while users with MMS-capable phones each send over 93 SMS messages per month, they send less than one-half of an MMS.[6]

---

[6] 'The State of MMS', Mobile Media, Baskerville Telecoms.

Several factors have contributed to the lower than expected uptake of MMS, not least the interoperability between handsets with different screen sizes, image resolutions and audio formats, as well as the price differential between SMS and MMS. However, we believe there is a more fundamental reason, caused by a basic misunderstanding of how customers think. Norway's Telenor is one of the leading operators in MMS with 70 % of its customers owning MMS-capable handsets and with each user sending an average of between five and 10 MMS messages each month.[7] Yet these same users are still sending around 80 SMS messages per month.[8] This clearly indicates that MMS is not *replacing* SMS – it *complements* it.

2.2.6.1   What we can learn: customers do not think like us

By 'us' we mean those who work in the mobile communications industry. We conceive, design, implement and launch mobile services; we understand how they work and we value the technology that goes into their design. We assume that, if we think a service will be useful, everyone will start to use it.

However, all too often we do not truly understand what drives people to use mobile services. Customers do not think like us. They do not care about the technology *per se*, but how it improves their lives. The assumption by some that MMS will rapidly replace SMS as a messaging medium is flawed and betrays a fundamental misunderstanding of customers and the way they use technology.

SMS is a fairly crude service, but it allows people to quickly and cheaply satisfy very important practical and emotional needs:

- the practical need of sending timely or urgent information – *will b l8 home; mtg cancelled; meet u in the pub @6*;
- and the emotional need to maintain relationships – *how ru?; thx for last nite; happy birthday!*

MMS will not replace SMS in the majority of these situations simply because text is adequate to convey the message. In fact, when asked many young people are unable to describe the typical content of their SMS messages, saying it is just 'junk'. This indicates that the act of sending and receiving messages is of more importance than their content. Increasing the richness of the content in this case is thus irrelevant.

To create success with MMS we need to determine situations when sending a picture is the natural thing to do. Here are three circumstances where a picture really does paint, if not a thousand words, at least more than 160 characters:

(1) *When it significantly enhances the information or emotion to be conveyed* – our experience shows that the most common subjects for MMS messages are people's children and pets, indicating the strong emotional impact provided by a picture of a loved one.
(2) *When it provides a simpler or cheaper alternative to a traditional means of communication* – traffic figures released by Swedish operator TeliaSonera show that more

---

[7] Source: Yankee Group, 'Multimedia messaging service begins to deliver on its promises', January 2005.
[8] Source: EMC.

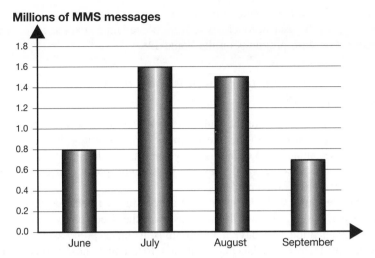

Figure 2.3 TeliaSonera Sweden, MMS traffic by month, 2004[9].

MMS messages are sent during the summer months of July and August (Figure 2.3). The obvious conclusion is that people on their summer holidays are using MMS as a more convenient, more personal and potentially cheaper substitute for the traditional postcard.

(3) *When it creates an entirely new service opportunity* – a popular and growing trend is the 'mobile blog' or 'moblog', (a mobile derivative of a 'web log' or 'blog'). A moblog allows people to create a personal diary on the Internet to which they upload pictures and personal information that can then be shared with friends and family. MMS provides an ideal tool to capture events in people's lives and have them instantly available on a public arena. Now anyone can be famous.

2.2.7 Mobile music and mobile TV – the final frontier?

Two of the most recent mobile applications to create a media buzz are mobile music and mobile TV. Music download services have been launched by many mobile operators in conjunction with their 3G networks, which make it relatively quick to download a full music track to the mobile phone. At the same time mobile TV services, such as Sprint's MobiTV, have appeared offering a range of news, sport and entertainment channels from established TV providers, which can be streamed to mobile phones in real time.

It is too early to judge whether mobile music and mobile TV have the potential to reach mass-market status or whether they will remain niche applications. When it comes to the customer experience mobile music and mobile TV must be treated differently. The mobile phone offers CD-quality music and is an alternative to other handheld music

---

[9] Source: Yankee Group Wireless/Mobile Europe, 'Multimedia messaging service begins to deliver on its promises', January 2005.

devices, whereas TV on the mobile is not intended to replace traditional TV but is *complementary*. It keeps the user in touch away from home and can act as an interactive tool used in combination with programs on the home TV.

2.2.7.1   What we can learn: from now it is about persuading customers to use applications

Mobile music and mobile TV are the latest and probably the last media to be mobilized. We – or at least those of us with the latest phones – can now talk, listen, read and write text, hear tones, see images, take photos, play games, watch videos, and now listen to music and watch TV, all on a single device. The mobile phone has caught up with all 'traditional' media, albeit on a smaller scale.

However mobile music and TV will not be successful just because they use the latest cool technology, it will be because the services meet the needs of customers. In fact this is true of any application. We have a full toolbox to make pretty well any application we choose on any form of media, but let us not forget that most people still only use their phones for voice and SMS, and unless we give people good reason to use new applications this situation may continue.

So let us spend a chapter getting to know our customers and learn what makes them tick. Armed with this knowledge it will be easier to design applications that appeal to customers and which will persuade them to part with their money.

2.3   A CHANGE TO A MARKET-LED APPROACH

The mobile industry is gradually changing. We have started to accept that a technology-led approach does not work with mobile applications, and that it is not sufficient simply to develop an application and launch it on the market with a brief promotion campaign in order for it to enjoy rapid growth.

Mobile operators, traditionally technically dominated, are increasingly allowing their marketing departments to determine the applications and services that are launched to the market. And this is right – after all it is the marketing department's role to understand the market and the customers therein.

Many people have the mistaken belief that marketing simply involves the production of promotional material. Others confuse 'marketing' and 'sales' – often allocating both roles to a single person in a company – where in reality these functions are very different.

In a nutshell, marketing is concerned with matching a company's applications and services with the needs of the market. A key premise of good marketing is that it is much easier to design successful applications if we have some understanding of the people who are likely to use them. Marketing is thus a cyclical process that starts and ends with customers: first we have to understand the market in order to design the applications; then we have to promote the applications in order to sell them to the market (Figure 2.4).

(1) Market understanding is acquired through research and analysis of customers and potential customers in the market (as well as the competitive situation).
(2) With this information we identify and develop the applications that will attract specific customer groups and that will best compete in the market.

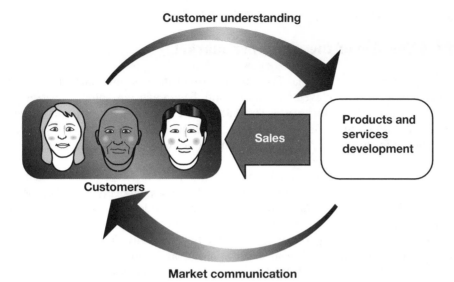

Figure 2.4 The marketing cycle.

(3) Market communications then create the awareness and demand in the market for the applications, focusing especially on the customer groups identified in the research.
(4) Finally, the sales process persuades people to buy our applications.

So whereas many companies involve the marketing department only in the latter stages of development prior to launch (market communication), the most critical marketing (market understanding) ought to be performed before development starts, in fact as an input to the initial definition phase.

To develop an application with no understanding of customers and the market is effectively to work in the dark, with the chances of success based only on luck and the instincts of a few individuals. Conversely, developing an application even with a comprehensive understanding of customers does not guarantee that it will be successful, but it improves the likelihood of success and, most importantly, provides a benchmark that can be used to compare the actual results with expectations, from which modifications to the application or the way it is marketed can be made in an informed way.

It is also important to note that, as well as understanding customers *before* we develop an application, it is crucial that we monitor the market following its launch in order to spot when something is not working, gain an understanding of why customers are not responding as expected, and take rapid steps to improve or remove it. In Chapter 10 we look at how customer monitoring and feedback can be used to improve our service offerings.

Let us continue this chapter, though, by taking a closer look at our customers themselves. We are going to analyze customers, both consumers and businesses, from a theoretical standpoint in order to gain insight into their underlying behavior. This will allow us to identify the people most likely to use mobile applications and the type of applications that interest them.

2.4   UNDERSTANDING CONSUMERS

2.4.1   The size of the consumer market

Consumers are individual people. The ultimate size of the consumer market is the entire population of a country, or even the whole world. Few products or product classes, however, are adopted by the whole world. The 'addressable' market, i.e. the maximum number of people for whom the product is relevant, is generally less than the total population. The challenge for marketers is to capture as much of the addressable market as possible, in as short a time as possible.

The size of this addressable market varies widely for different products, and many factors determine the rate at which the market adopts new products, including people's specific needs and interests, the number of alternative products available and of course the willingness and ability of people to pay the price tag.

For mobile applications, our addressable market is physically limited to the number of mobile phone users, and, for more advanced applications, by the number of handsets with the requisite features (GPRS, MMS, Java, etc.). There are things we can do to maximize the addressable market, such as promoting them so that people are aware they exist and understand their benefits, and ensuring people's phones are configured, making it easy for people to access the applications. The applications themselves may persuade people to upgrade their phones, thus expanding the size of the addressable market.

However, the ultimate size of our addressable market depends very much on the applications themselves and how many people see them as relevant to their lives. This will vary greatly from person to person. Some people love playing games on their phones; some want to obtain news and other information services; others see the phone mainly as a means of security and like the idea of an emergency locator facility.

In order to simplify the task of designing the right mobile applications for the right people, we need to divide the market according to the interests and needs of customers; we need to *segment* the market.

2.4.2   Consumer segmentation

2.4.2.1   Why segment consumers?

It is impossible to interview and understand every individual in a market, so we have to group people in order to make the research manageable (and affordable!). Equally it is impossible to satisfy everyone in a market with a single application. Segmentation involves defining groups of consumers with similar characteristics for whom we can create different applications to suit each group. In some cases it may even be possible to target different customer groups with the same application simply by packaging and promoting it in different ways.

2.4.2.2   Ways of segmenting consumers

Consumers can be segmented in numerous ways: by age, sex, nationality, occupation, income, social class, personal interests, etc. The aim is to define segments that can be clearly distinguished from each other, but where the individuals in each segment share certain common characteristics.

The parameters used to segment customers depend heavily on the industry in which we operate. The market for toothbrushes or breakfast cereal is very different from that of mobile phones or DVD players, which is different again from that for cars or garden furniture. A person's weight may be an important factor in defining market segments for a breakfast cereal, but is not relevant when it comes to DVD players, where a person's income may be more relevant. The challenge for us is how to define segments relevant for mobile applications.

Segmentation models are most powerful when they can be used over a long period, when people's behavior can be compared from year to year, and long-term trends can be identified. Now telecommunications is an incredibly fast-moving industry and it is impractical to design a segmentation model that keeps pace with this change. If we were to adjust the model every time a new technical innovation appeared, first we would spend a fortune repeatedly conducting customer surveys to update research data, and more importantly it would be impossible to compare one survey with another because the base criteria for the model would have changed in between.

The segmentation model we will now describe is one developed by Ericsson in its customer research division Consumer and Enterprise Lab. Founded in 1995, Ericsson's Consumer and Enterprise Lab performs continual consumer and business research in representative telecommunications markets throughout the world.

The result of this research has determined that, while change in the telecoms industry is rapid, people's basic values and attitudes to change remain fairly constant – if someone believes technology is a good thing today, they are likely to continue believing this for years to come as new products are introduced into the market. As a result, the segmentation model we shall describe uses as its base people's basic values and attitudes to technology and technological change, which gives the model the long-term stability we require.

2.4.2.3   The six driving forces of telecoms

By analyzing the results of interviews with 30 000 telecom consumers in 30 countries, performed annually over a 6-year period, Ericsson's Consumer and Enterprise Lab has identified six driving forces that determine and shape people's values and attitudes to technology, telecommunications and the market for mobile applications. The six forces are:

- connectivity;
- innovation;
- social awareness;
- stimulation;
- social status;
- tradition.

Connectivity

Connectivity is the most basic of functions provided by the mobile phone and the need to stay in touch is a fundamental part of most people's lives. The desire for connectivity is particularly strong in technologically advanced countries, driven by people's increasingly

fast-paced and stressful lifestyles, the difficulties of meeting people in increasingly anony-
mous societies, the need for security, and the need to maintain long-term relationships
with family and friends. Of the people surveyed by Ericsson, around 60 % scored connectivity highly as a driving force, with more women – notably between 20 and 39 years
old – represented than men.

Innovation

Technology innovation is generally perceived to be a positive force for world economies
and for the lives of ordinary people. Globally between a quarter and a third of people
regard innovation as a driving force in their lives, the figure being higher in developed
countries such as the USA and Japan.

Consumers who are motivated by innovation tend to be well-educated, young, active
people with the means to afford new products. They are the trendsetters and are often
the first to adopt new mobile products and services. They act as the opinion leaders in a
market to whom others look to for direction.

Social awareness

Socially aware people are concerned more with social issues than fulfilling their own
individual needs. Something like 40 % of consumers globally rate social awareness highly,
with women, parents and people over 40 over-represented.

Social awareness as a driving force has grown steadily in recent years, driven by
concerns over globalization and climate change, the exposure of corporate scandals such
as Enron and Worldcom, and an increasing mistrust of politicians.

Socially aware people have average mobile usage and they see technology as a necessary part of their daily lives, but only where it provides a clear benefit. They are not the
earliest adopters of technology but the public listens to their opinions, hence they play an
important role in the validation of new products and services for the mass market.

Stimulation

Stimulation as a driving force is diametrically opposed to social awareness. People driven
by stimulation are self-oriented and desire instant gratification. They seek out new experiences, adventure and excitement as a way of distracting them from what they regard as
the dullness of everyday life.

In the mobile world this group is a small but very important segment of the population.
In the Consumer and Enterprise Lab global survey, only around one in seven people score
highly on stimulation, consisting predominantly of young (half are under 29 years of age),
mostly single people.

They are big users of technology, spend a lot of time online and are heavy users of
portable devices such as MP3 players. Eighty percent own a mobile phone, which they use
much more than other consumers, both for personal conversations and nonvoice services
such as downloading music, sending images and playing games.

Social status

These consumers have a particular need to be recognized as socially and financially
successful. Prevalence of this behavior varies widely; globally 15 percent of people are

classified as social status seekers but this can rise to up to 35 percent, particularly in developing societies.

Ownership of mobile phones by this group is average, but those that do own them are medium to high spenders and use their phones more frequently and for longer periods than average users. They have a high interest in mobile applications, particularly those that are visible and that incorporate recognized brands. However, these people are not early adopters and they often lack technical know-how.

Tradition

Not everyone regards new technology in a positive light. As much as 25 percent of the global population believes new technology causes more problems than it solves. This attitude is notably strong in developing markets and in countries where religion plays a prominent role. The mobile phone, however, is generally regarded in a positive light as it provides people with an important communication link that provides a sense of safety and security.

Of people for whom tradition is a driving force, around half possess a mobile phone, which they use significantly less than the average mobile user. These people are often price-sensitive and rarely use all the features of the phone.

2.4.2.4   A consumer segmentation model for telecoms – Ericsson's Take Five

Building on the six driving forces, Ericsson has constructed a consumer segmentation model called Take Five.

The MarketReality™ Monitor

To describe the model let us start with what we refer to as the MarketReality™ Monitor. This takes the form of a graph with two axes, shown in Figure 2.5.

The vertical axis records consumers' attitudes to change. Towards the top of the Monitor are found people who are open to exploring new ideas, who look to the future and are enthusiastic about high-tech products. At the bottom are people who are content with things as they are (or were) and prefer to maintain a safe and stable life. They regard most new technology as unnecessary or even harmful.

The horizontal axis deals with people's personal values and their relationships to others, as well as the reasons why they adopt technology. To the right are people who are very self-oriented, live for the moment and seek new experiences for their own instant gratification. Technology for these people is a source of fun and entertainment. In contrast, people to the left are more concerned with the greater good, whether for their families and friends, society in general or the environment. They appreciate new technology but only if it provides a clear lasting benefit to themselves or others.

We can see that the driving forces described in Section 2.4.2.3 can be mapped neatly onto the Monitor, as illustrated in Figure 2.6.

The three driving forces of Innovation, Stimulation and Social Awareness apply to people more open to technological change; the forces of Tradition and Social Status reflect people who prefer stability. The forces of Stimulation and Social Status describe people who are self-oriented and who seek instant gratification; Social Awareness and

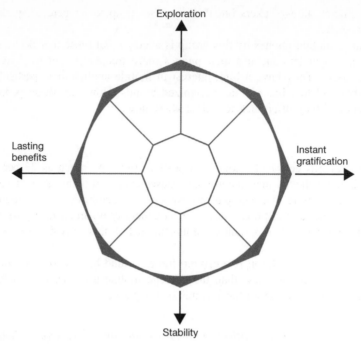

Figure 2.5 The MarketReality™ Monitor.

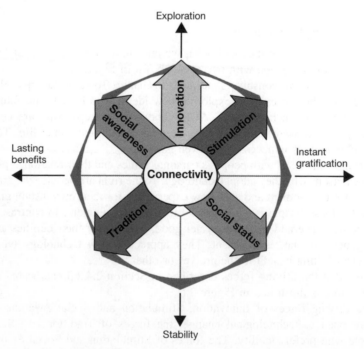

Figure 2.6 The six driving forces mapped onto the Monitor.

Tradition are the forces that focus on lasting benefits. Connectivity is fairly neutral in this model and is thus placed in the center.

From this it is now a short step to defining our base consumer segments.

The five consumer segments

The Take Five model defines five basic consumer segments (hence the name):

- Pioneers;
- Materialists;
- Sociables;
- Achievers;
- Traditionalists.

Each segment has clearly defined characteristics that distinguish it from the other segments and that can be measured in consumer surveys. Figure 2.7 shows the segments placed on the MarketReality™ Monitor.

Pioneers are, by definition, first adopters and are interested in anything new, including the latest high-tech mobile phones. They will experiment with the technology and provide inspiration to others.

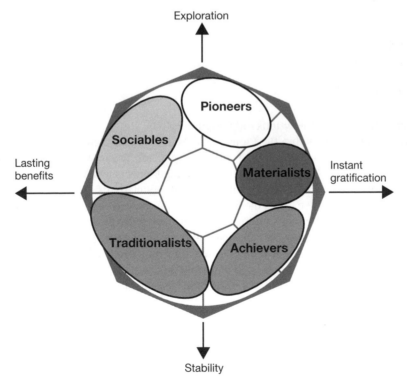

Figure 2.7 The Take Five segments.

Materialists are self-oriented but are less open to exploration and so tend to be early followers. 'Fun' is the thing that drives them; ringtones, icons and games are popular with this segment.

Sociables have a positive attitude to technology and new mobile services, provided their usefulness and benefit is clearly demonstrated. They are therefore also early followers.

Achievers, like Materialists, are self-oriented but prefer more traditional status symbols such as powerful cars and designer clothes. They may own the latest mobile phone but will use it mostly to impress friends and will probably not use most of its features.

Traditionalists, like their name suggests, are the last segment to adopt new services. For them the mobile phone provides a measure of security and is used only when absolutely necessary.

Life-stages

We shall now subdivide the five basic segments according to age, education and employment status, which will allow us to drill down deeper into the specific behaviors of different consumer groups. With considerable analysis, not repeated here, this gives us eight subsegments, as follows:

- Young Pioneers (P1);
- Adult Pioneers (P2);
- Young Materialists (M1);
- Adult Materialists (M2);
- Educated Sociables (S1);
- Older Sociables (S2);
- Educated Achievers (A1);
- Older Achievers (A2);
- Traditionalists (T).

Our segmentation model now looks like Figure 2.8.

As the figure shows, within the Pioneer and Materialist segments there is a significant difference between young people and adults. In general Young Materialists are notably more exploratory than Adult Materialists, and Young Pioneers and Materialists are slightly more driven by instant gratification than their adult counterparts.

Distinct subgroups also exist in the Sociables and Achievers segments, but here it is not so much age but education and employment status that is the distinguishing factor. Sociables and Achievers who are well educated and/or in managerial jobs are, as a rule, more exploratory and more self-oriented than those with a more basic education or employment (here 'educated' means with a college/university education, or equivalent). Appendix 1 details the key characteristics of each Take Five subsegment.

The size of the Take Five segments

Figure 2.9 shows the relative size of each Take Five subsegment in different countries. As can be seen, there are variations. For instance in Spain 23 % of the population are classed as Materialists (M1/M2) whereas in the UK Materialists make up only 15 % of the population. The differences between countries are due to many factors, including

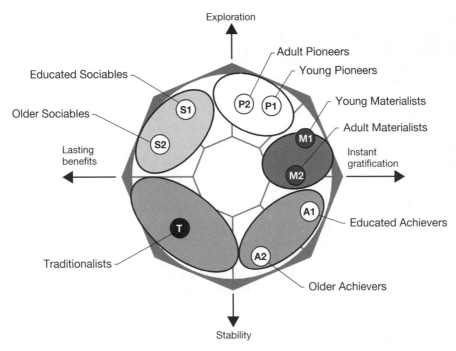

Figure 2.8 Take Five subsegments incorporating life stages.

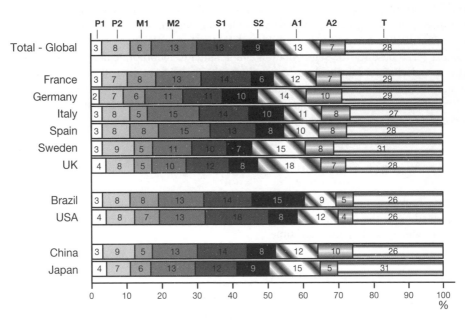

	P1	P2	M1	M2	S1	S2	A1	A2	T
Total - Global	3	8	6	13	13	9	13	7	28
France	3	7	8	13	14	6	12	7	29
Germany	2	7	6	11	11	10	14	10	29
Italy	3	8	5	15	14	10	11	8	27
Spain	3	8	8	15	13	8	10	8	28
Sweden	3	9	5	11	10	7	15	8	31
UK	4	8	5	10	12	8	18	7	28
Brazil	3	8	8	13	14	15	9	5	26
USA	4	8	7	13	18	8	12	4	26
China	3	9	5	13	14	8	12	10	26
Japan	4	7	6	13	12	9	15	5	31

Figure 2.9 The sizes of the Take Five segments in different countries.[10]

---

[10] Source: Ericsson Consumer and Enterprise Lab, Global Facts and Figures Report 2004.

cultural attitudes, the countries' economic status, the level of technological development, particularly with regards to mobile and Internet penetration, and so on.

Despite these differences the principles of the Take Five model and the validity of the consumer subsegments described have been proven worldwide in countries with widely different economies and cultures.

Note that we do not claim Take Five is the only consumer segmentation model appropriate for the telecommunications industry, nor is it so rigid that it cannot be adapted for use by companies who already have a segmentation model of their own. Here it simply provides us with a tool to define clearly differentiated consumer segments, which will make it easier to specify and develop applications appropriate for different sectors of the population.

Case study

In 2004 a major GSM operator in the Asia Pacific region requested Ericsson's assistance to gain a deeper understanding of its customers in preparation for the launch of its 3G network. Hitherto a predominantly technology-led company, senior executives in the operator recognized the need to acquire concrete data on customer needs and behavior in order to produce a meaningful roadmap for the following 1–5 years.

To this end a consultant from the Consumer and Enterprise Lab was assigned to the operator for 6 months working as part of the operator's Strategic Planning Division, and with strong support from the company's key strategic stakeholders.

The core of the assignment was a detailed total market survey to identify the size, needs and behavior of different consumer groups within the whole population. The result was an optimized market segmentation platform reflecting current and future consumer behavior and opportunities across fixed, mobile and Internet technologies. The Take Five model was used as a base for the segmentation and adapted to incorporate additional dimensions such as individuals' family status, e.g. with/without children, as well as being a platform to analyze specific needs such as small businesses or different ethnic groups.

The operator is now recommending that a similar exercise be performed for subsidiaries beyond its local country operation.

2.4.3   Market adoption

We have looked at how to divide up the mobile market into segments, and we have a fair understanding of the characteristics of the people who make up these segments. Now we shall introduce the dimension of time.

2.4.3.1   The market adoption curve

The market adoption of any product or service follows a classic S-curve: a slow start followed by accelerated growth to reach mass-market penetration, and finally a tailing off as the market saturates. However, it is not enough just to launch a product and wait for the world to come and buy it. It is critical to understand that growth does not happen randomly, based on who happens to walk past our shop window on any given day.

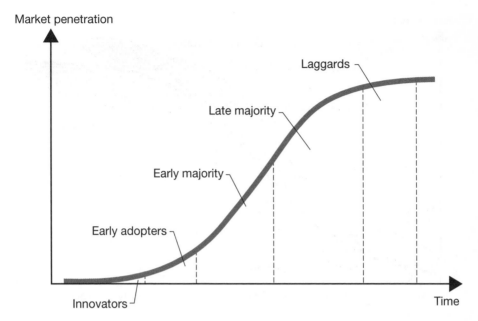

Figure 2.10 Market adoption curve.[11]

Products are adopted in sequence by the various customer segments in turn, as described by Everett Rogers in his 1962 book *Diffusion of Innovations* and illustrated in Figure 2.10.

The first people to adopt are called the innovators, followed closely by the early adopters. The early and late majority segments represent the breakthrough into the mass market, and market saturation occurs with the laggards.

A new product or service will not be taken up by, for example, the early majority without it first penetrating the innovators and early adopters. The people in each segment in effect 'sell' the product or service to the following segments. This has important consequences for how we design and launch new applications, as we must ensure that we first target and gain acceptance from the innovators and early adopters in order for our product to have any chance of penetrating the wider market. Conversely we may not need to directly target the mass market at all, as these people will learn of the products from the earlier segments.

2.4.3.2 Take Five on the S-curve

Figure 2.11 shows the Take Five segments placed on the market adoption curve. Note that this is just an illustration; the relative segment sizes and rates of adoption vary from product to product and from market to market. In fact it is under our control to manage this adoption, a topic we will take up in Chapter 5.

---

[11] Everett M. Rogers, *Diffusion of Innovations*, 1962. (Current edition Simon & Schuster International, 2003.)

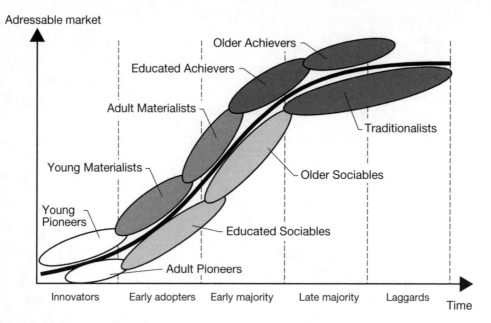

Figure 2.11 The Take Five segments mapped onto the market adoption curve.

As their name suggests, the Pioneers, both Young and Adult, are the first to adopt new technology, followed closely by the Young Materialists and Educated Sociables. The other segments are classed as early or late followers and adopt products based on the recommendations of the early adopter segments. The Young Materialists influence the Adult Materialists who in turn influence the two Achiever segments. Similarly the Educated Sociables provide the main influence for the Older Sociables, who finally bring in the Traditionalists.

2.4.4   Applications for early adopters

It is clear now that, in order to give our new applications the best start in life, they must be designed to appeal to the innovators and early adopters in the market (P2, P1, M1 and S1 in the Take Five model). The question now is, what sort of applications are these people likely to be interested in?

2.4.4.1 Adult Pioneers (P2)

Adult pioneers are fascinated with technology for its own sake. They frequently own more than one mobile phone and, if you are unfortunate enough to be dragged into conversation with one, they will describe in detail all the phones' features as well as the relative advantages and disadvantages of each. Adult Pioneers will try out any new application just to see how it works, but they are more likely to regularly use applications that enable them to organize and automate their lives, such as the ability to connect to the

Internet wirelessly and synchronize their phone or Personal digital assistant (PDA) to their PC in order to maintain up-to-date calendars and schedules. They like to have constant access to up-to-date news, sport and financial information. The BlackBerry device would be particularly attractive to this segment, both because it is cool technology and because it serves a useful purpose.

It is very important that we impress these people, as their opinions will determine whether other people try out our applications.

2.4.4.2 Young Pioneers and Materialists (P1, M1)

We have grouped these two segments as their needs and interests are similar; Young Materialists just tend to adopt services a little later than Young Pioneers. If we think back to the beginning of this chapter, we can see that the success of SMS, ringtones and logos is largely due to their adoption by these two segments, both of which have a need for fun, to demonstrate their individuality and maintain strong ties within their peer groups. Research also shows a strong interest among these segments in most new and fun services, including music downloads, games, content downloads, picture messaging, sports and news information, video trailers, and so on.

2.4.4.3 Educated Sociables (S1)

The mobile needs of the Educated Sociable are not dissimilar from those of Adult Pioneers – they are also interested in tools that improve efficiency and help organize their busy lives. They are also attracted to image-based applications such as photo messaging and video telephony that enhance their contact with family and friends. However, whereas Adult Pioneers enjoy mastering new technology, Sociables have a higher need for technology to be well designed and easy to use. The technical difficulties associated with many mobile applications, for instance MMS, has meant that mobile applications have not yet penetrated this segment to any great degree. If we contrast this with i-mode in Japan, which was explicitly designed to be easy to use, penetration of the Sociables segment would have occurred far more easily, which perhaps explains why i-mode rapidly achieved mass-market adoption.

2.5 UNDERSTANDING THE BUSINESS MARKET

2.5.1 High potential, slow growth

The potential market for mobile business applications is huge. The mobile phone, laptop PC and PDA, together with wireless data networks have now made it possible for many employees to perform their work outside the traditional workplace whilst keeping in touch with colleagues and company resources.

Mobile solutions for the business market include mobile office applications such as mobile e-mail/personal information management (PIM) and remote Internet/intranet access; applications to automation sales forces and field-support staff; mobile banking and finance; fleet management; and enterprise-wide solutions such as customer relationship management and supply chain management. Despite the potential, the penetration of mobile applications into the business market has been as slow as, if not slower than, with consumers. Fewer than one in 10 mobile workers use their mobile phone for anything other than the basic functions of voice and SMS.[12]

A major inhibiting factor in the growth of mobile business applications has been the separation between the traditional suppliers of enterprise solutions – IT developers, systems integrators and value-added resellers – and the suppliers of mobile connectivity – mobile operators. The former lack knowledge and experience in mobility; the latter have little or no experience of complex enterprise systems and solutions, rarely have direct sales channels to enterprises and do not speak the language of the IT/IS manager, who is often the key decision-maker in how a company spends its technology budget. To overcome this, some operators have acquired or partnered with system integrators and enterprise IT providers in order to offer total solution packages for mobile business.

Another factor behind the low penetration of mobile business applications is simply that most businesses are quite conservative in their spending behavior. Before they invest in new technology, companies need to see a clear business benefit in a new application, such as increased productivity and efficiency, reduced costs or improved customer satisfaction.

Therefore, it is even more important to understand business customers than it is with consumers. Not only must we understand the needs of companies and employees, we must also be able to demonstrate explicitly how our applications will create a significant improvement in the bottom line.

2.5.2 The size of the business market

Figure 2.12 shows that 75 % of companies in the USA and 93 % of European companies consist of fewer than nine employees (the so-called SoHo – Small office/Home office). Most of the remainder employ less than 50 people. Proportionally there are very few corporations with over 250 employees, but these employ over 50 % of US workers and 35 % of European workers.

When it comes to 'mobile' workers, i.e. employees whose jobs require them to move within or outside the company, the picture is different again. Research reveals that in Western Europe there are 52 million mobile workers, of which only 4 % work in companies with over 500 employees; 82 % work for companies with fewer than 100 employees (Figure 2.13).[13]

These contrasting pictures present something of a dilemma for mobile applications providers – should they target the largely untapped corporate market, focus on small to medium companies where the majority of current mobile workers are, or go for the SoHos that make up the majority of companies? The answer of course is it depends – on the applications they have to offer, the employees they intend to target and the industry segment in which the companies operate.

---

[12] Source: Ericsson Consumer and Enterprise Lab.
[13] Source: Decision Tree Consulting, 2004.

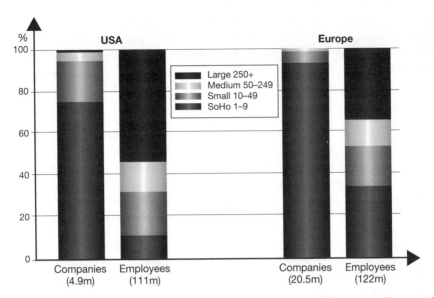

Figure 2.12 Distribution of companies and employees in Europe and US (source: Eurostat, 2002; US Census, 1999).

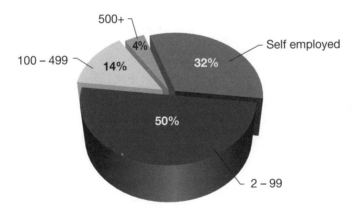

Figure 2.13 Distribution of mobile workers in Western Europe (source: Decision Tree Consulting).

2.5.3   Segmenting the business market

There is such a diversity of applications, industries, companies and working roles that it is impossible to define generic 'early adopter' business segments – as we did with consumers – which will take our applications and promote them to a wide market. Some applications may be adopted quickly by certain companies but totally ignored by others; equally the same application could be useful to multiple industries and companies of widely different sizes, but may require different marketing, distribution and support strategies for each.

The key, as always, is to start with the customer, identify the *unmet* needs of companies and employees, and then develop solutions to solve their specific problems. So let us look at the business market from different aspects and consider the most appropriate approach in each case.

2.5.3.1 Company size

SoHos, with their few employees, must be treated almost as consumers, with solutions promoted through consumer media and available in retail stores or online. At the other extreme, many large companies look to capitalize on their existing investments in back-office systems and choose to adapt and extend their existing products and systems to introduce mobility. Application suppliers must therefore be familiar with legacy systems and be able to adapt and integrate their applications to them. Smaller providers should look to partner with IT service providers, systems integrators and/or value-added resellers with whom many large companies will already have a relationship.

Some large enterprises, with a diverse mix of communication systems and devices, with little or no interoperability or compatibility between them, may decide to start from scratch and look for a total, customized solution that simplifies and streamlines their communications end-to-end, including the provision of mobility. This market is increasingly being served by the major suppliers of back-office systems, including IBM, SAP, Oracle and Microsoft, who now include mobility as an integral part of their solutions.

Small to medium-sized companies experience many of the challenges of larger organizations, but they do not necessarily have the budget for large, bespoke solutions. They seek instead scalable and flexible solutions that can be quickly and simply installed, with the ability to be customized to their specific needs.

The key with any company, regardless of size, is to reach the key decision makers: the IT or telecoms manager whose budget it is likely to be impacted and who will be responsible for implementing the application or solution; the finance manager who will approve the solution based on its business benefits; perhaps the manager of the department(s) most affected by the application; and ultimately the CEO.

2.5.3.2 Employee roles

Mobile applications require mobile workers. There is little point trying to sell a mobile solution to a company whose entire workforce sits at a desk the whole day (for example a call center). An important factor in segmenting the business market is therefore to study the roles of workers and the degree to which they are mobile.

Figure 2.14 shows various work groups, each of which is a cluster of occupations with similar work patterns, levels of mobility and usage of communications tools such as Internet, fixed and mobile phones and PDAs.

As can be seen, in the upper part of the diagram there are five groups that involve a medium to high degree of mobility. (Appendix 2 describes the typical characteristics of people in these roles.) However, we must be careful not to confuse mobility with the need for mobile communications. For instance, many in the construction industry may be classed as 'super mobile blue collar workers', and are highly mobile, but actually communicate very little in their jobs.

Degree of mobility

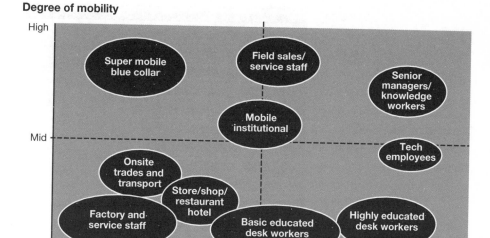

Figure 2.14 Degrees of mobility for different work roles (source: Ericsson Consumer and Enterprise Lab, 2004).

Another critical factor is who pays for employees' phone bills. In general it is rare for nonmanagerial staff to have company-paid mobile phones. Senior management and knowledge workers are the most likely groups to have their phone bills paid for them, and some tech employees and field sales/service employees receive support from the company, but even within these groups the vast majority of employees pay their own bills. Clearly when employees are expected to pick up the tab, the likelihood of their voluntarily adopting mobile applications is diminished.

The groups most likely to adopt mobile applications are in fact those that are already using communication tools in their work, namely senior managers, knowledge workers, tech employees and, to a degree, field sales/service staff.

The results of a global survey into the needs of the working population shows that the applications working people are most interested in are those that allow them to maintain seamless communications with colleagues and office tools while away from the office (Figure 2.15). These include voice forwarding, single phone number, mobile e-mail, calendar synchronization, corporate directory and access to the company Intranet and corporate databases.

2.5.3.3 Industry segments

Finally, we can consider so-called 'vertical applications' that address the needs of a specific industry segment, such as finance, health, transport, retail, etc. Research by Ericsson as well as by ARC Group has identified a number of industry segments that have been early adopters of mobile solutions.[14] Let us look briefly at each of these.

---

[14] ARC Group, 'Mobile enterprise – from mobile office to integrated solution, worldwide market analysis and strategic outlook 2004–2009'.

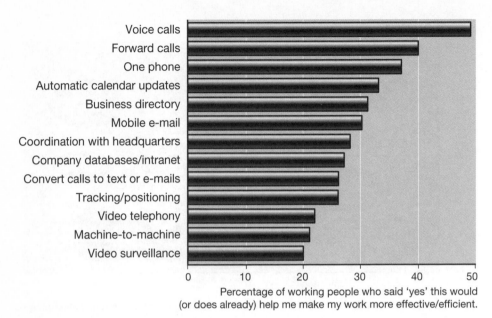

Figure 2.15 The applications considered most useful to working people.

Transport and distribution

This is an industry that is mobile by its nature; tasks that are enhanced by mobile applications include accurate tracking of goods, status reporting (collection/delivery), issuing of new jobs, route management and navigation, and remote surveillance of goods and vehicles.

Utilities

Utility companies involved in water, electricity and gas provision have a wide geographic distribution and employ many mobile workers. Mobile tools make the issuing of jobs and status reporting more efficient; the introduction of a remote meter via wireless links increases efficiency and reduces employee costs.

Business and professional services

This industry segment includes consultants, computer/IT staff and other specialists. Often working away from the office with clients or at home, they have a high use of mobile office solutions, collaboration tools such as video conferencing and online meetings, as well as remote access to corporate data and research.

Financial services

This industry is already a high user of IT and communication systems. Many banks have introduced mobile banking and trading, either via SMS or WAP, and insurance agents are able to fill in claims while with clients and send them back to base for rapid processing.

Security is a key factor in application design – both to prevent fraudulent access and to maintain confidentiality of client data.

Healthcare

The healthcare profession offers many opportunities for mobile solutions, both intra-facility within hospitals and healthcare centers and for remote employees such as doctors on call and emergency personnel. Applications include basic messaging and appointment scheduling, tools to reference and maintain patient records in real time, access to medical directories and information, stock control, diagnostic aids, and so on.

Retail

The retail industry uses mobile devices and applications for real-time sales monitoring and inventory control, supply chain management and sales force automation. Use of mobile phones as discount vouchers is not yet widespread but could provide a powerful link between loyalty schemes, retail offers and promotional campaigns, whereby registered customers receive regular product offers pushed as a message to their mobile phone, which then acts as a discount coupon.

2.6 SUMMARY

With a few notable exceptions, the mobile applications business has been driven by technological development rather than led by customer need. It is easy to praise the foresight of those who have succeeded and point the finger at 'obvious' failures but, as we all know, we work in a young, fast-moving industry where taking risks is not only inevitable, it is *necessary* to succeed. In retrospect there have been few real failures in mobile applications, but success in many cases could have been achieved far quicker than was actually the case. We cannot turn the clock back but we can hopefully avoid repeating the mistakes of the past.

There is no 'killer application', in fact there are many – each of us has our own killer application, but it is as impractical to design an application specifically for each individual as it is impossible to design a single application that satisfies an entire population. We have to break the problem down into manageable proportions. We do not need to be experts in human psychology to do this, we simply need a few tools to make the job easier. The aim of this chapter has been to provide these tools and to give some insights into who uses mobile applications and why. Armed with this knowledge we can now move on to the task of creating a winning service offering.

3

Creating a Winning Service Offering

3.1 EXPLORING SERVICE CREATION

The evolution of the mobile into a media channel contrasts two very different approaches to service creation. On the one hand are the rigid control processes required to secure new and often untested end-to-end technology that is used in telecoms and on the other is the fast moving channel approach of TV and other media industries where the underlying technology is of limited importance and everything is about creating attractive and appealing content at a reasonable cost.

What approach is the most efficient and when is it applicable – trial and error vs a well-documented service creation process secured in a formal chain of milestone approvals? Should services be created based on technology push or based on what the market 'wants'? Can we know what the market wants? Is being first to market most important, or should we learn from the mistakes of competitors and launch as late followers? Is there any logic behind how the telecoms operators have traditionally dealt with service creation or is it only a legacy?

In this chapter we explore some different approaches to understand the rationale behind various dimensions of creating new services. This helps us analyze the conditions under which we have the best fit with these different approaches to achieve a winning service offering.

3.2 THE THREE DIMENSIONS OF SERVICE CREATION

Mobile media applications are different from media in more established formats such as TV and radio. A specific challenge with telecoms is that the value lies in the network effects, neatly expressed by Robert Metcalf's law: the 'value' or 'power' of a network

Mobile Media and Applications – From Concept to Cash: Successful Service Creation and Launch   Christoffer Andersson,
Daniel Freeman, Ian James, Andy Johnston, Staffan Ljung
© 2006 John Wiley & Sons, Ltd

increases in proportion to the square of the number of users (or nodes) on the network. What it means in plain English is that there is very little value in a network service when there is no one else to call and that the perceived value increases very quickly the more users are connected to the network. When launching new services like video telephony, messaging services or even plain voice digital mobile telephony 15 years ago, this has posed some great challenges. The initial users are likely to see little value in the new technology. What is also interesting is that users have more creativity than designers! Users find different ways to use the service than those it was originally designed for! Another challenge when launching media services with the mobile phone as a channel is that telecoms technology, and the digital distribution standards it is based on, is still immature. This means that there are both risks and opportunities with the handling and integration of the emerging and evolving technology enablers. Yet we have to dare to be innovative, creating and putting new services on the market that are built on new concepts and technology. This technology push dimension of service creation we call *integration of technology enablers*, see Figure 3.1.

Another challenge that perhaps is not so industry specific is how to translate, identify and satisfy customer needs with creative service offerings that meet the demands of our customers. We also have to do this at a price that the customer is prepared to pay and with the timing that is within the window of opportunity in the market place. This is all about *translating customer demands* to a service offering that has attractive and economical feasible market conditions. Often technology is used in a different way than first expected. SMS was not originally designed to serve for flirting, chatting, goal service updates and many more creative person-to-person and content-to-person applications. The users, in particular the youth-dominated early-adopter segment, showed the needs and demands on the technology. A good service creation methodology hence needs to be able to drive and evolve the technology based on an efficient translation of customer demands into technology requirements. By doing this better and quicker, competitive advantages can be created for any player involved in the service creation chain.

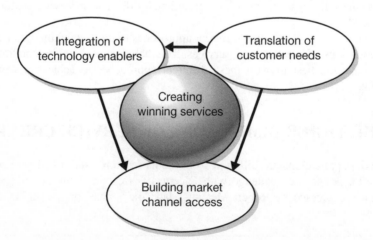

Figure 3.1 The three main challenges that have to be dealt with in creating winning service offerings.

Over time the mobile device plays an increasingly important role as a media channel. However, the experience is increasingly built as sensations created – and shared – across many media: TV, the mobile network or the Internet. Hence the convergence aspect is important, as is building access that is seamless to the user. Building market channel access becomes an integrated part of service creation – this is the third and last component in our service creation model. When leveraging the mobile as a media channel, new constellations of media and access providers have to be realized to create a winning service offering. How to build the market channel access for emerging opportunities based on both the successful translation of customer needs and the integration of technology enablers is explored in the last cornerstone of the three-dimensional service creation model.

These dimensions are not mutually exclusive – rather, a new service should maximize as many as possible of these three dimensions. However as the challenges and risks are somewhat different, it is worthwhile breaking down and analyzing the service creation in these three dimensions to optimize the service offering. You will therefore see in the examples that each service creation approach has a little of the others, but the emphasis and risks are dealt with differently.

3.3   TECHNOLOGY PUSH – CREATING SERVICES BASED ON INTEGRATION OF TECHNOLOGY ENABLERS

By combining and integrating technology enablers, new creative services can be put on the market space. To make this more concrete let us look at a couple of examples: the traffic navigation service Wayfinder and the multipurpose enabler IMS (go back to Chapter 1 for a definition of IMS). We will use these examples to illustrate important aspects in the creation of service offerings that are generic to other applications and useful for a media player that wants to understand some of the fundamentals underpinning service creation in telecoms. We will also relate back to these examples in the following sections on service creation based on customer demand, 'market pull' and market channel access.

Wayfinder is a traffic navigation service whereby the user can see where he or she is and get up-to-date information on traffic congestion, rerouting suggestions, and best route to the destination. Figure 3.2 shows a screenshot of the service where the route is highlighted and clickable references to hotels and restaurants are highlighted on the map. What are the constituting technology enablers and what are the risks and opportunities in integrating technology enablers into the solution? The service combines a smart-phone's capability to run an application that displays the position of the mobile device on a vector-based map. The position is obtained by transmitting the coordinates from a global positioning satellite (GPS) receiver that is connected to the phone using Bluetooth. Information about the vector map and current traffic situation in the area is transferred using either GPRS or WCDMA (wideband code division multiple access) as a bearer connecting the mobile device to a server that has a database with all relevant information. Wayfinder is an example of a service creation chain that relies heavily on technology enabler integration (even though it, of course, does this with the ultimate goal of satisfying customer needs).

One could think of many ways to integrate additional technology enablers to make this service more attractive and compelling (of which some have been implemented already). For example, using enhanced position information from the network rather than using

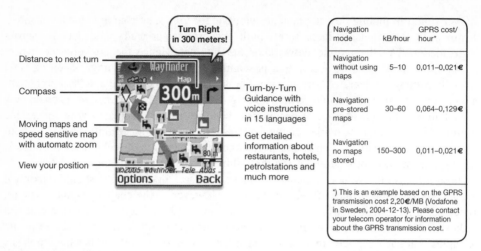

Figure 3.2 Wayfinder is an example of a creative service combining many technology enablers. Source: Wayfinder.

the GPS receiver – this would make the end user equipment less expensive and easier to use. Using 3G instead of 2G as a bearer would enable higher speed, which would allow for more detailed maps. Using Java 2 Mobile Edition (J2ME) and standard application programming interfaces (APIs) rather than a native application would allow for a broader base of mobile devices, but would perhaps limit the functionality. Enhanced display functionality such as mobile flash could allow for media presentations on tourist attractions along the road that the user can access by clicking on the interactive map displayed in the mobile device. The list of possibilities is endless but let us have a look at the individual parts to understand the challenges and how realistic they are over time. We can see the enabling parts in three main areas: mobile devices, service network and access network.

There are a number of differences to take into consideration when analyzing these solution components. It is worthwhile to look at the *dynamics of change* of different technology enablers.

The *access network* is the mobile operator's largest capital expenditure by far. The access network is typically some 80 % of the total network investment. This is because extensive coverage requires a large network of radio base stations. To tune and build the coverage to reach the service accessibility of today's GSM network required billions of Euros in investment and many years of constant refinement. Adding new bearer technology such as WCDMA or wireless local area network (WLAN) will hence take considerable time. The core network is where the control logic is lying, the intelligence of the network. The service network is a diverse set of enablers like location, streaming and messaging servers. The *core* and the *access network* are the parts that are to the largest extent driven by standards. This means that there is a long lead-time before new core and access network functionality is available across all networks. The core and access network have a longer lead-time than the specific servers used to feed the traffic information and the maps. This part of the service network is more of a traditional 'IT', system with building components that can be changed and developed more dynamically than the access network as long as it does not imply a change in the interface to the access network.

The *mobile device*, in this case a smartphone with Bluetooth capability and a large display to show the maps, has a relative slow pace of change that involves the air interface and interaction with the core network and service network, since it requires extensive interoperability testing and coordination of standards. However the phone capabilities in terms of memory available and screen quality have a fast development pace. Another factor is the volume of the addressable market: if we decide to use the latest device features this limits the target market. We will review the implication of the unsynchronized development between the mobile device and the networks further in Chapter 6.

The conclusion of this discussion is that these three parts and their related standards set the boundaries of what we can do. For the pragmatic applications developer it is better to develop its application within these boundaries rather than to wait for the promise of new technologies – it takes time to go from press release to implementation across the target networks.

Let us now look at IMS as second example to show how these boundaries are pushed over time. Perhaps the easiest way to grasp IMS is to say that it makes it possible to communicate with the mobile phone in a similar way to existing popular Internet networking applications like Yahoo Messaging, MSN messaging, ICQ or Skype. With IMS we can see in our mobile address book who is available, busy, happy, sad or whatever we choose to show as our online-status. IMS makes it possible, just like Skype and other Internet networking applications, to have videoconferences, exchange files, chat and talk. For the users of these networking applications, IMS will just be a natural extension to enable on the mobile the same things they are used to doing online. For others it will be seen as a new way of communicating. Instead of first choosing how and to whom we wish to communicate, much more will be driven by what happens online. We can shift between different means of communication depending how the conversation evolves: maybe we start chatting with a happy fellow who is online, then when chatting becomes too slow we 'lift the phone' and start talking and end up sharing some video shots from the last party we went to.

Technically speaking, IMS gives the possibility to maintain session control independently of access method. I can be online and be accessed using both the Internet and the mobile phone. It is the foundation for multiple services and defines the standard interfaces to be used by service developers. For the application developer the promise is more reuse of functionality across different networks and the possibility for new creative mutiaccess applications. IMS uses the Session Initiation Protocol (SIP) and other standardized protocols, which provide functionality such as presence information (e.g. see in the address book that your friend is on the phone), multimedia session handling (e.g. playing a game while talking) and adaptation of multimedia capabilities (e.g. showing low-resolution pictures on the mobile and high-resolution pictures on the computer). Based on the promise of IMS we could envisage myriad services. Exciting? Yes, but wait a bit … did we not hear the same hype for MMS and 3G? Well here is the dilemma of the technology push dimension of service creation. Metcalf's law told us that, initially, when there were few connected users, the value would be close to zero. Another paradox is that the initial technology both enables too many *possible* different services and limits us to a few *plausible* service propositions due to poor end-to-end integration. At the same time we receive little guidance from users as to what they want. Our goal has to be to reduce the time to gain network effects which we know create value to users. To do this we seek to replicate known, existing behaviors to the mobile, that is, creating propositions that are easy to communicate to and test with customers. That makes it possible to evaluate what enabler integration is required to deliver the solution and drive the requirement for development in the direction that users point out to us.

In the technology push dimension, we evaluate the technology end-to-end in light of different customer propositions we want to invest in. This gives us both a requirement for change and warning flags for barriers to overcome. The possibilities of the technology inspire new service propositions that we bring to further evaluation in the tool gate funneling process described in the customer demand dimension.

To make the discussion more concrete let us look at four potential services that can be realized with IMS:

- *Push to talk* – this is a walkie-talkie type of service where one user can start talking to another by pushing a button on the mobile. The service is delivered over the data network (GPRS, WCDMA) and allows, in its simpler format, only one user to talk at a time.
- *Interactive gaming* – this is a service that allows two or more users to play against each other interactively in a game like poker, chess or quake. While you could play multiplayer games before, now you can talk while playing (synchronized) and more easily connect to friends who would like to play.
- *Instant voice messaging* – Today you can leave someone a voicemail when they are busy or unavailable, but there is no way to see what their status is. IMS enables a user to leave a message directly in the voice mail if the receiving party is busy. This is made possible by knowing the status of users, who can identify themselves as available, busy, etc.
- *Share image and motion* – this allows users to exchange images and videos during a conversation.

In summary, IMS will enable new ways of communicate rather than replicating the way we already communicate over the Internet by enriching the media format. It will enhance the network effects since the interaction is extended to other media and network types than just the mobile phone by using protocols (SIP) that are widely accepted across many industries. This is highlighted in Figure 3.3.

These services can also be combined. For example one could think of a use case where two players start a push-to-talk session while they are engaged in a combat game, see Figure 3.4!

There are clearly different ways to implement the mentioned services. We will here contrast two different enabler integrations to illustrate the impacts on the service offering. One way is to build specific end-to-end solutions for each service. This has been seen as a success factor in Japan and Korea, where the operators decided centrally what to launch and then dictated the specification at a very detailed level for suppliers of mobile devices, service networks, core networks, etc. Take as an example a chess game where games connect to a game server, or a push-to-talk solution that is transferring voice messages based on a proprietary mobile device client–server interaction. This vertical way of building a solution, mostly with components developed only for one specific application, we will call the 'stove pipe' approach. This we contrast to the standards that IMS incorporates to allow for end-to-end solutions of all kinds of applications from chess

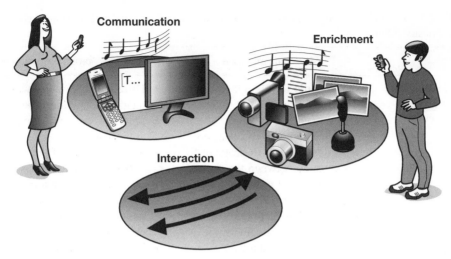

Figure 3.3 IMS has the potential to enable many different services.

Use-case >> check Budd-list/presence >> start a text chat >> invite parties to an online/network game >> communicate while playing the game

Battle over the real estate

Different service entry points

Initiators vs recievers' freedom of controlling the communication

How does the presence of information affect the way we communicate?

Service start from Address Book or service icon/button

Move from one session to another

Available games present for parties

PTT built in to the game

What possibilities should be built in, 'all combinations' vs optimization?

Figure 3.4 The multiparty gaming use case gives rise to several usability and enabler integration questions.

games to push-to-talk. In Table 3.1 we summarize the two different enabler integrations: IMS vs stove pipe. The advantage of going with the proprietary solution is generally time to market, and that the specific application is richer than the standard allows for. This could gain valuable early customer insights, but risks getting a bad reputation in the eyes of the important ambassadors, the early adopters, if we implement it when the technology is not ready. The benefit of adhering to standards is that services in the long run will work across networks and between wider ranges of terminals. We can also avoid expensive changes as most proprietary solutions eventually migrate towards the standard.

Table 3.1 Example of enabler integration analysis contrasting using standard IMS components with a proprietary solution.

	IMS 'standard based'	Stove pipe 'proprietarysolution'
General enabler integration aspects	• Standards for charging, provisioning and operations and maintenance, data exchange between operators • Builds on the SIP protocol for exchange between the mobile device and the functionality in the core network • Risk of longer time to market due to slow agreement process of standards. Challenging to get the whole industry to adhere to and interpret standards	• Different/proprietary ways of handling signaling, routing, presence, authentication and authorization • Many applications in a network = many nodes • Dedicated integration towards operation and maintenance, charging, provisioning per service • Each application (Share; PTT etc.) requires a dedicated association between the mobile device and network • New interconnect agreement required for each application (with each operator)
Push-to-talk	• OMA-PoC/PAG compliant • Multiple handsets, multiple brands, multiple target groups – all interworking • Network-to-network interoperability • Reuse of standardized presence/GLMS • Use cases, signaling and protocols well defined • Future service enhancements and IOT (test fests, etc.) part of standardization process • Terminal requirements: 2.5G terminal including PoC client, dedicated PTT button, Hi-Audio speaker	• Not OMA-PoC/PAG compliant • Limited choice of handsets from system supplier • Limited or no interoperability with other terminals and other networks • Proprietary presence • Use of cases, signaling and protocols partly proprietary • Limited possibilities for future service enhancements • Significant effort for interoperability testing outside of standardization • Support of proprietary phones required for long time after launch • Terminal requirements: 2.5G terminal including PoC client, dedicated PTT button, Hi-Audio speaker

Table 3.1 (*continued*)

	IMS 'standard based'	Stove pipe 'proprietarysolution'
Share image and motion	• Instantaneous sharing of images and video during ongoing CS voice call • Compliant to 3GPP work item 'combinational services' • Interoperability among terminals and networks • Standard finished by end 2005 (standardization 3GPP Release 7), early drop proposed • Whole application family • Not limited to images and video, but extendable to everything being sent via MSRP/RTP • Future service enhancements and IOT (test fests, etc.) part of standardization process • Terminal requirements: 3G terminal including camera • IMS capabilities • Share client according to the standardization body 3GPP	• Based on MMS extension – store and forward • Share image: picture MMS • Share motion: video MMS • Different user experience • Improved terminal GUI required to offer easy MMS access during phone call • No live video, but recording, sending and downloading of video clips • Reuse of existing network infrastructure, capacity expansion required • Terminal requirements: 3G phone including camera, MMS capable • Improved terminal GUI
Instant voice messaging	• Leaving people direct voice messages depending on presence status • Extension of OMA-PoC Service • Candidate for OMA-PoC R2 • Reuse of presence and GLMS • Seamless integration in PoC client • Interoperability among terminals and networks • Interworking towards non-PoC phones (e.g. MMS) • Standard to be finished by end 2005	• Based on MMS extension – store and forward • Sending Audio MMS • Different user experience • Improved terminal GUI required to offer easy audio MMS creation • No integration in PTT, presence, GLMS • Reuse of existing network infrastructure, capacity expansion required • Terminal requirements: 2.5G phone including IVM button, MMS capable

(*continued overleaf*)

Table 3.1 (*continued*)

	IMS 'standard based'	Stove pipe 'proprietarysolution'
	• Future service enhancements and IOT (test fests, etc.) part of standardization process • Terminal requirements: 2.5G terminal including PTT/IVM button, Hi-Audio • IMS capabilities (SIP, SDP, MSRP, RTP, etc.) • OMA-PoC R2 client included	• Improved terminal GUI
	• Peer-to-peer gaming platform based on IMS • Use IMS/SIP for session set-up, session control, charging • Reuse of presence and GLMS • Push-to-talk can be used while playing • Minimal additional standardization required • Terminal requirements: 2.5G terminal (3G/EDGE for higher bandwidth) • IMS capabilities (SIP, SDP, MSRP, RTP, etc.) • SIP game runtime environment (J2ME/Mophun) • OMA-PoC client (for simultaneous PTT)	• Proprietary gaming platform • No person-to-person reachability based on presence status, but 'login' to 'gaming rooms' • No reuse of presence and GLMS • Probably additional SMS/voice conversation required to 'invite' game partners • No voice/PTT/IM communication while playing • No interoperability with other gaming solutions, no interworking with other operators • Terminal requirements: 2.5G phone (3G/EDGE for higher bandwidth) • Game runtime environment (J2ME/Mophun)

IOT, interoperability testing. IVM, instant voice mail. PTT, push-to-talk. GUI, graphical uses interface. OMA, open mobile alliance. PoC, push-to-talkover cellular. 3GPP. Third Generation Partnership Program.

To find out what this implies in the technology push dimension, we have to cover at least the following aspects:

(1) User – mobile device; *user client.*
(2) Mobile device – access network capability; *device access.*
(3) Implementation of service network/core network (residing in the core network and/or the service network); *service network access.*

When we look at the user client integration there are a couple of key questions. What is the availability of phones? How interoperable is the solution on different handsets. A stove pipe solution may give short-term gains but limit the network effect long-term to one vendor's phones or only to high-end niche phones. This question is more tricky than it first seems because the more configurations and set-ups and clicks are necessary to access a service, the less revenue-generating it becomes. Experience suggests that anything but off-the shelf integration means that we can only count on early adopters as a realistic addressable market; this means that we risk to losing as much as 90 % of the market until the issue is resolved. The implication has a direct bearing on the economical feasibility of satisfying the customer needs. This is exactly what needs to be analyzed in the technology push dimension: what are the consequences of different enabler integration? What does it translate to in usability and service uptake? When can services be launched? How do we create the network effects?

Many data services depend on, for example, the adoption of head-sets to allow the customer to watch the screen while talking, which impacts the service suitability. One factor that has made data services more popular in Japan and Korea than elsewhere is the higher penetration of clamshell phones, which allow for larger screens, which in return allows the user to watch and type simultaneously more efficiently. More clicks and complicated set-ups directly translate into less service revenue.

Let us next look at the device access, in other words the integration between the device and the access network. These interfaces are handled through standardizations and specification of the network release. Even though network vendors differentiate with some unique functionality, the important functionalities, like in this case IMS, will eventually evolve into standards. Standardizations and time frames for implementation have, as we saw in the discussion of dynamic change, slow cycles (years).

- Are there any dependencies to new network releases? When is it realistic to expect the release to be in place? What can we do with what is in place already?
- How are the standards likely to evolve? If it is built on proprietary technology, will it work across all mobile devices?
- How is the service provisioned? Can it be activated easily or without modification of the system?

As we have learned, access network represents a large investment and the pace of gaining ubiquitous access, including international roaming and countryside coverage, takes time if we are to count on new standards. The third level we use is the service network aspects. The key is then to see how technology components map to possible service offerings. To do this we use a technology roadmap that shows how technical requirements develop over time and deliver service functionality. It is important to map the key constituting parts of the end-to-end system to get a view of when it is really realistic to have 'co-ordinated' availability between mobile devices, applications, enablers, core infrastructure and access infrastructure.

In Figure 3.5 we have outlined what such a technology roadmap could look like for IMS.

In conclusion what we see is that enabler availability is constantly developing and differs across customers and networks. The development of enablers follows a different logic and pace for different parts. Network access technology (GSM, WCDMA, CDMA), service network enablers (MMS, SMS, positioning) functionality and mobile devices do not have coordinated release cycles, which causes considerable fragmentation and end-to-end gaps that have to be dealt with in the service development process. In the technology enabler integration we preferably start with established customer behaviors and services and enhance this with an enriched communication experience. In more mature mobile services, such as SMS, the technology is a less important success factor (since it is a given), but for newer services this factor has to be weighted in. Some enablers enable

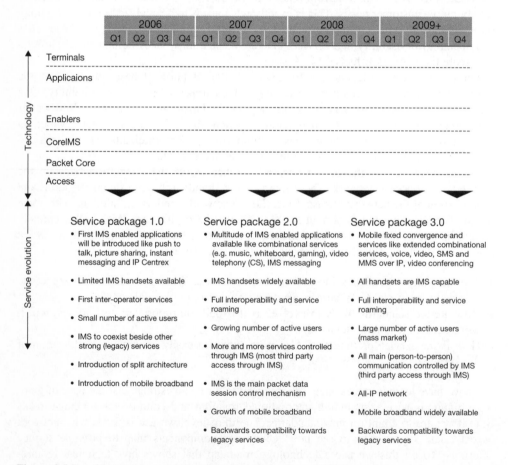

Figure 3.5 Resulting service evolution based on technology roadmap.

many different combinational services (and an enrichment of existing services). This poses different challenges to the service development project that require an integrated market pull and 'technology push' perspective.

3.4   MARKET PULL – CREATING SERVICES BASED ON CUSTOMER DEMAND

Let us continue with the second bubble in our methodology for winning services – creating service based on customer demand. As we recall from Figure 3.1 there is a double arrow between the integration of enabler technology and translation of customer needs. We can create a winning service starting in any of these dimensions, but we have to reiterate and adapt based on our findings in the other dimensions. Ultimately any service has to be based on customer demand but with the enabler integration dimension we make sure that the service is feasible and seek inspiration for satisfying demands with better enabler integration.

Any supplier of technology enablers has of course created its enabler (even though it is not always obvious!) to allow for creation of services that satisfy customers' needs directly or indirectly. However, frequently it is built based on analogies with existing successful services in a different context, i.e. fixed voice–mobile voice, Internet–wireless Internet, fixed broadband–mobile broadband, chatting over internet–mobile instant messaging, etc. What makes them fly or flop? The challenge of translating a customer demand for a mobile media application gained from an insight in a customer need is far from straightforward as we have seen in Chapter 2. This is about how to remove barriers and how complexity can be reduced to simple graspable concepts that trigger user interest. Can we build on established customer behaviors and notions from the Internet world (like instant messaging)? Can we simplify the user interface (like the iPod where hundreds of man-years went into designing a simple interface)? If we look at the example of IMS, we want to use concepts that are familiar at least to the early adopter segments, something that is a more intuitive, natural way of communicating, see Figure 3.6.

How does the enabler integration impact this? How attractive are the new services? How do we control the risk and financial suitability of our service creation projects?

In the customer demand dimension we also include the notion of launching the service at a price that is economically feasible. It becomes a program management challenge to optimize the key dimensions time, quality and economic resources.

3.4.1   Planning and controlling the service creation process

A key challenge in a successful service creation is to keep the two dimensions of enabler integration (technical feasibility) and customer demand (commercial suitability) throughout the process without adding too many overheads and bureaucracy that kill the creativity and nimbleness in the project. They are dependent on each other, i.e. a different enabler integration changes the price and the attractiveness of the services, which in return results in a different customer demand.

Pre-IMS communication   **IMS communication**

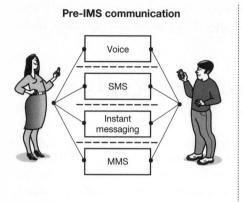

1) Deside on communication mode/media
2) Create content
3) Send/call the chosen person
4) Disconnect and reconnect if changing media

1) See who is available beforehand (presence) and how
2) Create session and select media
3) Change media and contenders in real time

Figure 3.6 How can we make the new services intuitive?.

The larger the scope of a new service offering, the more important it becomes to manage the creation process and break it down into smaller steps to keep risks at reasonable levels. This is why most operators have created a service development process that uses tollgates and designated steering boards to minimize investment in projects that are deemed less attractive, see Figure 3.7. Risk control and financial suitability have to be integrated elements of launching winning service offerings. We have seen many examples of operators who control this process in a detailed process description of several hundred pages – that no one follows. So the process itself can become a risk – keep it simple. Here we will focus on the high-level phases that exemplify where the important decisions and handovers happen. From an operator's perspective the challenge also becomes a change management issue. In terms of planning the service offering, methodology is required to select and value the most attractive service out of many possible ones, which we will illustrate in this section as well.

3.4.2 Service concepts and business case

Let us go back to the IMS example. One of the challenges in the early service concept phase is to quantify the revenue over time of different service ideas and feed that into the feasibility phase. As indicated in Figure 3.7, we go from idea generation and screening to a more detailed concept evaluation and feasibility study. A more detailed checklist and the key activities that have to be done is in this phase are given in Figure 3.8.

The elaboration of the business case is worth extra mention. In particular, for new services like IMS, user data will be nearly nonexistent. Benchmarks of other similar services can be used, but there will be many uncertainties as to how they will really

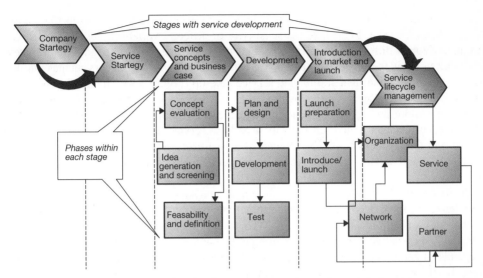

Figure 3.7 A phased tollgate service development process is used to minimize risk and spending in unattractive services.

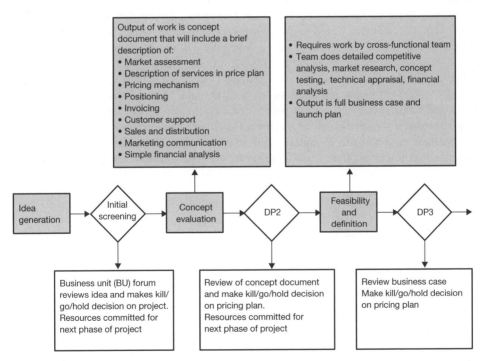

Figure 3.8 Summary of service concepts and business case phase with key activities and decision points.

translate to the service we want to launch. However the business case forces us to quantify and makes it possible to analyze the sensitivity of different variables. We are also forced to quantify how different variables impact each other. Take the IMS example: ultimately we want to know the revenue potential. We can link that to the key factors such as number of customers, terminal penetration and service take-up (Figure 3.9). The number of active users will increase over time, providing that the service is successful and manages to break into more and more segments beyond the early adopters. The adaptation models described in Chapter 2 give an indication of how service diffuses over time, which we can use in the business case. The number of active users depends on customers, terminal penetration and service take-up. Active users generate revenues by using and paying for the services.

In this example, based on a more thorough analysis of the two enabler integrations illustrated in Table 3.1, the absolute terminal penetration per service takes into account technology penetration [GSM/GPRS/UMTS (universal mobile telecommunications system)] and the relative terminal penetration per technology. The service take-up per service depends on end user experience and pricing. The terminal penetration is faster at the beginning for the proprietary solution since handsets were available earlier than in the standard IMS solution. However, the penetration into follower segments such as educated sociables and young materialists is slower due to the incompatibility issues. The improved end user experience increases uptake levels; interoperability between operators increases usage and service in the IMS-based solution. Over time, the absolute terminal penetration in the standard solution would surpass the proprietary one, which also contributes to a higher number of active users over a long time. We will not go through this in detail, but Figure 3.10 illustrates a typical output of such an analysis.

The concept evaluation is an iterative process to evaluate the attractiveness of services, identify as many relevant requirements as possible and, in particular, analyze those that really add value. More integration points and customizations of standard solutions cost time and money – even a small requirement for change can lead to a big cost implication. The business case helps us to quantify different scenarios. The bottom line value itself is normally less important than the possibility to simulate different scenarios and carry out

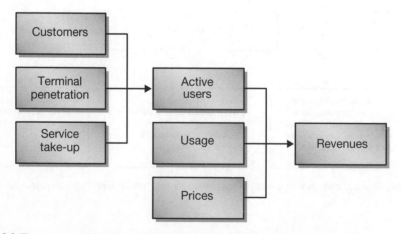

Figure 3.9 The revenue side of the business case and concept phase.

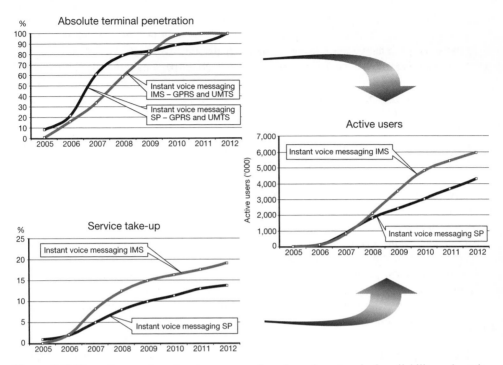

Figure 3.10 The active number of customers over time depends on terminal availability and service take-up.

sensitivity analysis. What are the most important factors? Which underlying assumptions have the biggest impact on the end result?

The so-called 5 M model[1] helps to evaluate the attractiveness of a mobile service and make sure that the full potential of all mobile traits is exploited. The framework is used to evaluate how much value a mobile service gives to the user and subsequently how profitable it would be in the supply dimension (for example application developer/technology provider/mobile operator):

- *Movement – escaping the fixed place*: the ability for a service to be operational as the user moves around. A service that has some locality aspect involved should adapt and benefit from the mobility aspect of the service. One can map the movement aspect with the previous analysis of enabler integration analysis. How well do the enablers deliver movement operationally? By what time can we expect it to be deployed nationwide, in the neighboring countries and internationally? When is the technology flawless enough to deliver moment aspects so that the service can attract other segments than the early adopters?
- *Moment – expanding time*: the power of the instant and the ability to manage time adapted to the situation we are in. In the 5 M model this also extends to managing

---

[1] Tomi T. Ahonen, Timo Kaspes and Sara Melkko, 3G Marketing, 2005.

tasks with more freedom in time, i.e. postponing tasks or catching up tasks thanks to the mobile service. It is also incorporates the notion of multitasking, i.e. sharing a video at the same time as we are talking to the other person.

- *Me – extending my community and myself*: this is about personalizing the service based on the specific need in a specific situation and place based on the user's history. This is perhaps the most important factor in the 5 M to consider. The more we can customize the content and the application behavior to each user's needs, the more successful it will be.
- *Money – expending financial resources*: the support of a mobile service to carry out financial transactions clearly contributes to the financial suitability. This also links to our confidence in the mobile service – the more we rely on a mobile service the more we are likely to execute financial transactions that are related to that mobile media or application. SMS voting is now something that is broadly adopted based on the relative high confidence in SMS and the low risk involved.
- *Machines – empowering devices*. The last M stands for the concept of enabling human–machine and machine–machine interaction (M2M). This can comprise extensions to existing services, such as sending an MMS to a server that prints a postcard that is mailed to the destination. It can be a Bluetooth-enabled device used to control a computer's PowerPoint presentation, or interaction with a parking machine. It can also be a lottery machine that is connected to and controlled from a central system remotely.

Another important test is time to market. Since time is money, the demand analysis needs to look at when it is possible to launch, reusing as much as possible of the existing enabler functionality. We can easily end up with a wish list that can only be satisfied in two releases ahead, which of course makes the payback times less interesting. Traditionally there has been a focus on time to market in the telecoms market, where operators strived to be first to market with new applications. However several 'first to market' launches of applications like GPRS and MMS did not bring any competitive advantage. This in combination with increased focus on maximizing the free cash-flow that the telecom crisis brought to the industry has caused a trend to instead creating customer propositions that have real value, under slogans such as 'the most intelligent way to market' rather than 'the first to market'. This launch intelligence drills down to both waiting until the technology is mature enough and avoiding perfecting the technological solutions for too long without real customer feedback. It is a balancing act between launching solutions that work end-to-end and acknowledging that it will not be perfect at once but has to improve in several iterations.

Let us return to the IMS example. What can services can we launch based on IMS enablers? We gave four groups of possible services: push-to-talk, interactive gaming, instant voice messaging and sharing images and videos. The next step is to evaluate this service ideas using focus groups. Such focus qualitative studies shows:

- interest in the different services per segment;
- triggers to start using the service;
- key barriers to overcome;
- user patterns, like who to use the services with, when are the services used, how often, etc.;

- willingness to pay;
- attractiveness per user segment.

We identify the current use of services, service use by handset type, access technologies used and volumes of usage across the week. In doing this we also have to seek and identify third parties for attractive content and applications to create a complete service offering. This helps us to prioritize and develop the technical requirements and quantify different scenarios in a business case. The resulting evaluations of each service as a service card are depicted in Figure 3.11. This analysis is the basis for taking an informed toll-gate decision to proceed to the next phase, the development phase.

3.4.3 Development

Design and development considerations are dealt with in later chapters, but their key elements from a creation and planning perspective are covered in Figure 3.12. The whole logic behind the phased approach hinges on the experience that, once you enter the development phase, the cost is a factor of 10–100 higher than that of the earlier phases and hence it is important to only enter with the projects with true potential.

Case study: What did TeliaSonera learn from their IMS pilot?[2]
IMS is a major enabling technology that is going to be utilized in a great number of session-based telecom services. However, as with all new technologies, there are many uncertainties involved: what are the suitable services enabled by this technology and where does the money come from?

The rationale for early testing is not only to ensure right services at an early stage but also to enable smooth integration into the legacy network. From a commercial standpoint the pilots serve to identify the relation of the new services to the existing offering, and to test the customer interest and needs of the new services.

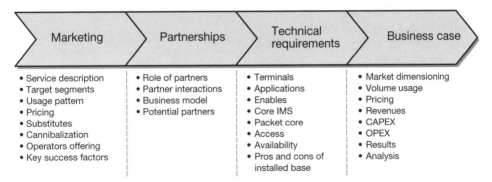

Figure 3.11 Sample service card of IMS service. (CAPEX, capital expenditure; OPEX, operational expenditure.)

---

[2] Source: TeliaSonera 2005.

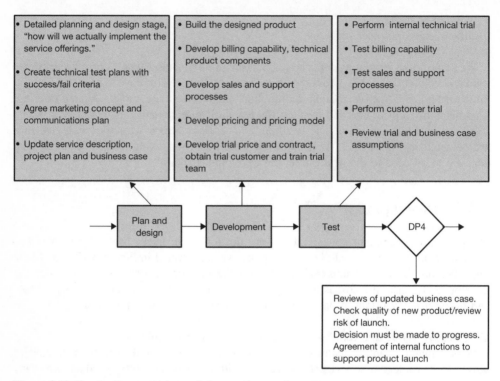

Figure 3.12 The development phase of the service creation process.

TeliaSonera started IMS end user piloting in 2003 with a multimedia exchange service called Media Community Application (MCA). Since then there have been a number of different pilots with various services and with several different user groups. The insight from these tests includes:

- Authentic end user feedback
- Feedback to application developer partners
- Technology testing
- Basis for service packaging and pricing
- Usage profiling
- Different delivery methods: preinstalled, OTA delivery, customer installed software
- Attractiveness of different applications: PoC, Multiplayer gaming, content sharing, etc.

The pilots have been used for the preparation for the IMS launch. The pilots have also given insight in different provisioning and charging methods.

The end user pilots have indicated a few mandatory enablers that are required for a successful service take-off. Among these is the interoperability of services between networks. Based on this TeliaSonera has taken an active role in GSM Association's interoperability and SIP trial work with a number of other operators to raise awareness of the maturity of SIP based services and to demonstrate candidate services.

3.4.4   Introduction and market launch

Even though we have tried to predict the service by comparative studies, focus group and trials, it can never compare to a real service launch. Any methodology for creating service inherently needs to incorporate early customer feedback. This is captured in Figure 3.7, and the 'introduction and service launch'. A brief checklist of activities that need to happen before launch is given in Figure 3.13, and a more detailed discussion is taken up in Chapter 9.

Subsequently continuous improvement is essential, which is captured in the model as the 'lifecycle management of the services'. We should bear in mind that many successful services are just new combinations of existing service elements repackaged and priced differently, marketed through new channels to new segments, improved by removing barriers, simplified technology, etc. For example, SMS was not an immediate success and neither was WAP, but now both these service enablers create good return of investment for many applications developers. This monitoring and continuous feedback is explored in three parts (Figure 3.14): the presentation of the commercial offering, decision subscription (or spontaneous use of the service) and service usage. We have to work with all these elements to make the market launch successful.

3.4.5   Service lifecycle management

The success of a mobile application throughout its lifecycle depends on the cost of attracting and acquiring new customers and how much revenue can be generated over time. This

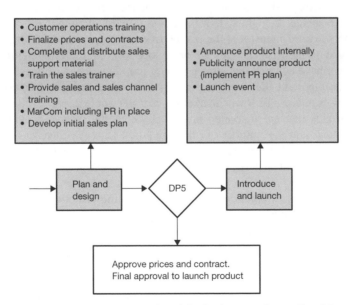

Figure 3.13 Key activities to prepare to launch and the final approval to go live. Marcom, marketing and communication.

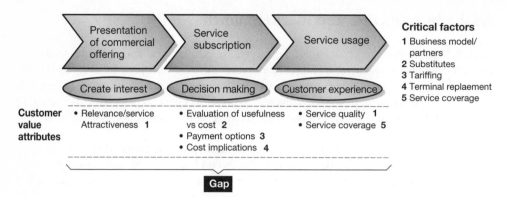

Figure 3.14 Introduction and market launch.

is not different from other industries, but there are some peculiarities and success cases that are worthwhile having a closer look at. A simplistic yet useful high level view of the product lifecycle is shown in Table 3.2, which is divided into four phases:

- attract;
- acquire;
- bond; and
- upsell.

For untested mobile applications based on relatively new enablers or device technologies, the grim reality is that the budget for attracting new customers is normally very limited. The advantage, though, is that the marginal cost of supply is typically close to zero. Like the software industries, the use of free trial copies and free trial periods has been a good way to increase service take-up at low cost. Optimus and Wind are example of operators that have adopted this strategy when introducing new services such as MMS and SMS. It is important that the test period is extensive enough to enable the user to establish a new behavior. To lower marketing cost per application, services are grouped into logical groups that are marketed on the portal of the service provider. The position of the content and the number of clicks required to access determines how easy it will be to attract users. It is becoming increasingly common in revenue sharing deals to define where the content and application should be placed given that this has a direct marketing value.

To convert an interested potential customer to a buying customer is of course key for a winning service offering. For services that require new customer behavior like photo messaging and video conferencing, it is important that the customer gets a chance to trial the service without cost. Trial periods and demo services in the shop have been shown to be successful in a number of campaigns. It is also worth mentioning the importance of the sales channel in creating uptake of new mobile applications. A comparative study between a large operator's own stores and independent distributors showed 20–30 % higher service adoption of multimedia messaging services to customers that had bought the MMS phone in the own stores. The reason is that the implemented sales process in the own stores

Table 3.2 Sample evaluation of a service attractiveness over its lifecycle

		Business communi- cations	Consumer communi- cations	eCommerce	Push technology
Attract	Competitive pricing	✓	✓		✓
	Image and brand name for quality	✓	✓	✓	✓
	Bandwidth and coverage	✓	✓		✓
	Content		✓	✓	✓
	Outsourcing/S1 capabilities	✓			
Acquire	Industry and product knowledge	✓			✓
	Service level agreement	✓	✓	✓	
	Customer education	✓	✓	✓	✓
	Free offer		✓		
Bond and upsell	Bundling services	✓	✓		✓
	Full service solutions	✓		✓	✓
	International presence	✓			
	Customer service excellence	✓	✓	✓	
	Reliability	✓		✓	
	Membership		✓		

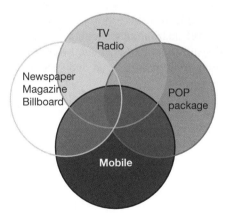

Figure 3.15 The mobile itself is an important tool to acquire new customers.

included a mandatory demonstration of how to send an MMS in the store. In budget-constrained campaigns, viral marketing and the mobile itself are important channels. It is one of the most impulsive, interactive and intuitive tools to maximize the effect of existing marketing (Figure 3.15).

3.5   BUILDING MARKET CHANNEL ACCESS

The mobile channel has some clear differentiators to other channels in that it enables *ad hoc* consumption of media service, interactivity and the identification and authentication of the user. The detailed information that can be gathered about the user's behaviors and preferences gives the opportunity to make fine-tuned segmentation and targeted offerings. Seeing the mobile channel as just one of multiple competing and complementary channels to address different audiences raises important questions: how do we make it an attractive channel that is easy to use for media companies and applications developers? Do the lower costs of digital distribution and production enable new bundles and ways of accessing the customers?

For the mobile operator the natural goal is to make the mobile channel as attractive as possible for media companies. To do this, clear interfaces and boundaries that define the 'creativity space' have to be established. The goal is to make it simple for the content owner to just 'add' the mobile channel to its existing distribution channels and reuse much of the existing processes. Figure 3.16 shows the typical content delivery process that can be used within the technical boarders of the 'creativity space'. The creativity space is where innovative content owners and application developers can use the mobile channel and still be certain that it will work end-to end, i.e. the triangle of mobile devices, service network and the access network provide proven integrated functionality. The creativity space defines clearly how the content supplier can package, distribute its content and be certain that the content gives a harmonized user experience across networks, devices and channels.

The mobile channel (and other digital distribution channels) removes barriers for production, packaging and distribution. The lower cost for distributing digital content enables more cross-selling opportunities. The Crazy Frog ringtone mentioned in Chapter 1 that became a best selling CD single with very low production and marketing cost in comparison to normal record label distributions exemplifies this. Since there are limitations of audio and visual quality in the mobile channel, production cost can be kept low and be compensated for by being more personalized and specific. We can choose the mobile

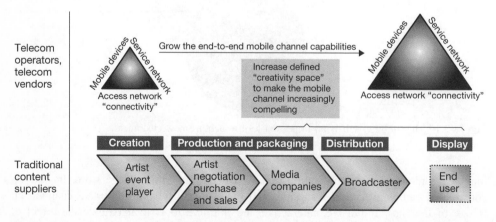

Figure 3.16 Increasing the attractivness of the mobile channel for content supply and consumption.

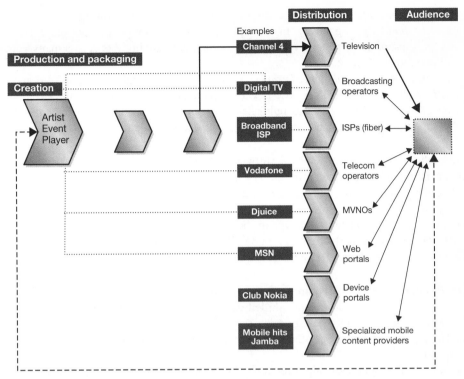

Figure 3.17 Digital distribution across many different channels. MVNO, mobile virtual network operator.

channel to be a primary marketing channel, whereas the revenues come from sales an advertisement in other channels. Alternatively, it can be a complementary secondary channel that provides highlights and interactivity to complement other channels. The more specific and direct feedback from the users in the mobile channel enables content providers to use this knowledge to impact the distribution in other channels.

Case study: How Terraplay helps game publishers to gain quicker market access[3]
Terraplay specializes in middleware that helps game publishers to distribute their games more easily across many different operators and many devices. On their customer list they have both mobile operators, like 3, Telefónica, Orange, TeliaSonera, Sony Computer Entertainment (PS2 titles), Motorola (handsets) and iTV (for an unnamed major TV producer), and content aggregators like Hiphop Mobile and Overloaded. This is a way to share IT infrastructure and integration costs related to providing the necessary gaming enablers. In particular Terraplay's platform Move does four things:

- *community building* – user tag, matchmaking, ranking, ladders and high scores, tournaments, buddy services, file and asset servers;

---

[3] Source: Terraplay.

- *community management* – online statistics, customer care, monitoring and alarms;
- *payment handling* – gateway to multiple payment systems, pay-per-play, content charge, subscriptions, generic on client, in particular resolving this across many games and operator networks;
- *multiplayer support and games kernel* – object switching, bandwidth optimization, protocol handling (UDP, HTTP, TCP). This is particular important since the experience from Mforma and other games publishers tells us that many multiplayer games have 80 percent higher downloads than the single player equivalent.

In addition Terraplay provides a software development kit (SDK) and development support is offered to the publisher and its games studio(s). In particular this SDK helps to resolve the porting issues and provide connectivity to the middleware (using Java, Brew or Morphun on the client side).

Let us for a moment put ourselves in the game publisher's seat. Say that we own the global intellectual property rights of James Bond and we want to launch a James Bond game (there is one already!). The process I follow is then:

(1) Write up a script or a requirement specification on what the game should look like.
(2) Send the specification to some skilled gaming studios (in-house or outsourced).
(3) Select a gaming studio based on the proposal.
(4) The gaming studio designs the game. This is when all graphic design and sounds are decided.
(5) Develop the game and ship a reference build supported by the 10–15 most common mobile devices.
(6) As games publisher we now want to distribute the game to reach the broadest possible audience.

The last point is where the headache starts. From the reference design I want to port it so that it supports maybe 60–70 devices. The game also has to support at least some six or seven languages, which means that we easily end up with several hundred 'skews' (variants). This game now has to be distributed through 50–60 mobile operators. They typically require the publisher to have all necessary game servers to support attractive multiplayer games. The players want to play with their buddies regardless of which operator they are using. This is where Terraplay comes in. They provide the enabling platform and connectivity to all mobile operators so that the publisher can focus on marketing and promotion. One issue, though, is that the revenue pie has to be split into yet another slice.

3.6   GOING FORWARD, HOW CAN WE CREATE MORE WINNING SERVICES?

There are innumerable potential products and services in the market but few highly successful ones. Choosing the optimal portfolio and getting early customer insight to shape the optimal product offering will be increasingly important. Quality is more important

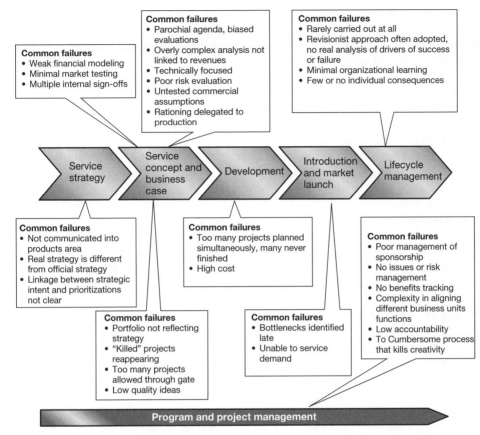

Figure 3.18 Overview of common mistakes in the service creation process. Source: Ericsson.

than quantity. To do this well any player has to work in partnership to create true end-to-end offerings. That requires that we have clear and simple development processes, standard interfaces and approaches. Once the service delivery mechanisms are established we should not try to impose rigid control processes that risk interfering with creativity. To establish a clear vision of what the product should look like and let this be a guide throughout the organization is easier said than done. One good case is Vodafone Live, where the service concept and service strategy are well linked to the later stages by clear customer propositions.

Case study: Vodafone live – proposition-based service creation

Vodafone Live is one of the largest scale projects ever undertaken by a mobile operator. While a lot has been written about Vodafone Live in general and about the portal, not so much has been written and analyzed from an inside point of view of how Live became a winning service offering. The concept is very much focused around mobile applications for consumers packaged under the Vodafone Live brand. The customer proposition is a selected and preconfigured handset that gives a one hot-button access to the Live portal, where services can be selected through an icon-based menu system. The success behind

Vodafone Live stems to a large extent from Vodafone 'muscle power', being the largest mobile operator and the massive multi-hundred-million euro branding investment, but nevertheless, here is some important knowledge that can be applied small-scale to create successful service offerings.

First is definitely the focus on the mobile terminals with preconfigured handsets, simplified navigation and consistent graphical user interface. This started with a close cooperation around the Sharp GX-10 handset (which was actually a GSM version of an existing Japanese handset but adapted and optimized for Live). To create an attractive handset at a reasonable price was one of the aims of bringing a Japanese handset to Europe. Apart from a tight integration with the Live portal the handset included enhanced Java for gaming and a Live-specific menu system.

A major achievement with Vodafone Live was the consistent look and feel across all of its operating companies. This was accomplished by starting with a clear definition of the look and feel of the handset, including detailed user scenarios. These user scenarios were agreed and stated in a master document shared across the whole implementation team. The propositions in the master document became the glue between the technical and commercial teams. They also provided an important linkage between each work-stream so that the focus on the customer propositions was maintained even in more detailed service definitions and technical evaluations. This linkage between the customer proposition and the implications for the technology, business and process aspects in the launch of Vodafone Live is shown in Figure 3.19. This overcame a common service creation issue: that roadmaps are defined based on technical enablers and that the service proposition and the technical solution are out of synch. This was handled effectively in the Live project by establishing the use cases and the look and feel early on in the project and then maintaining this linkage throughout the project with the help of the master document. The overall release planning was done based on customer propositions rather than enabler evolution. The solution at each operator within the Vodafone group was then tested against the defined customer proposition. Even though the underlying technology was somewhat different, the proposition had to be consistent across the group. These launch tests and follow-up quality tests were communicated and followed up by

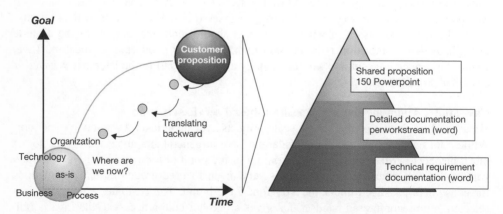

Figure 3.19 A shared customer proposition but different underlying technology.

Vodafone's top management, which created concrete benchmarks and best practices to develop Vodafone Live further.

However, by trying to mitigate all possible problems, barriers and pitfalls in the launch of new services, there is a risk of overcomplicating the process. The following quote from a BBC director on a telecoms conference on mobile TV, is worth reflecting on:

I don't know why you make it so complicated in telecoms, you spend weeks on developing business cases when you at the end of the day don't have a clue about the outcome. In the TV business we would allocate a budget to a skilled producer and then leave to him or her to prove that they can create a successful TV format.

We need to make it simple! The media world should not have to worry about the complex technologies that enable rich interactivity and improved mobile media quality. The phased approach makes sense for establishing the 'creativity space', but we should leave it for the media producers to be creative within that space – and collect feedback on how we can make it less limiting. One should hence clearly differentiate the service creation process between the implementation of enablers and the use of these enablers in the creative packaging of appealing content. The mobile channel competes with other channels for the eyes of the users, but can only do that by complementing the existing channels. The digital distribution and packaging barriers are lower and there are many parallel distribution media: digital TV, Internet protocol (IP) TV providers, web portals, mobile operator portal and device portals (like Club Nokia). Production costs must be kept low for mobile media given that the inferior broadcast quality of the mobile media instead has to be more tailored and specific to take advantage of the increased level of segmentation that the mobile channel allows.

3.7 SUMMARY

We have seen three dimensions of service creation: technology push of new enablers, market pull-driven translation of customer needs and seeing the mobile channel as one of many access channels that we need to combine in an intelligent way. For new telecoms enablers that span end-to-end, a phased implementation with a large degree of cross-functional control is still relevant. However, this should only be used to create a 'creativity space' for telecom media services. Once the boundaries of technology are implemented, so meaningful services can be launched, customer needs should drive how this space is enlarged and what constellations of media bundles appear within it. It is important that creativity is not killed by technical complexities and dependencies that require complex control processes.

3.8 FURTHER READING

Paul Golding, The Next Generation Mobile Applications, John Wiley & Sons, 2005.

4

Designing Services

4.1 DESIGNING SERVICES FOR SUCCESS IN THE REAL WORLD

Whilst it is now time to delve into more of the details around mobile application design, this chapter also provides a practical insight for non-techies into important aspects of application design that can have a strong effect on the commercial viability of a service. Ensuring that educated choices are made by business stakeholders and appropriate requirements are set by product managers at an early stage are key factors that ensure a successful service is what rolls 'out of the factory doors' at the time of launch. Conversely, requirements driven purely by the desire for advanced technology and unrealistic requirements of a solution can doom a service at the time of its conception. This chapter will provide a useful insight for all actors to prioritize the right design principles and provide a solid basis for product development decisions.

Many originally considered the development of mobile applications to be a simplified subset of software engineering. In most cases, it is now proving to be the opposite, as developers must contend with far more complex architectures and, especially on the device side, a less powerful, more diverse and often restrictive development environment. In addition, the complexity of designing applications that must function reliably in a wireless environment adds another challenge. This chapter will highlight a variety of design principles and strategies that exist in the mobile software development and media creation industries. The chapter will also cover the many considerations that must be made by developers as they design advanced mobile services.

4.2 SERVICES CLASSIFICATION

When setting out to design the next killer application, mobile device and network vendors will want you to believe that the whole world is open to you to develop new services. The opportunity to capitalize on everything from 7-megapixel cameras, advanced

Mobile Media and Applications – From Concept to Cash: Successful Service Creation and Launch   Christoffer Andersson, Daniel Freeman, Ian James, Andy Johnston, Staffan Ljung
© 2006 John Wiley & Sons, Ltd

3D graphics and gigabyte data storage on the device, to accurate mobile positioning and megabit download speeds makes it seem that the possibilities to create a ground-breaking application are infinite. Indeed, this is the case, but the realities of device feature penetration *in your target user segment*, which relates to the addressable market of the service (as discussed in Chapter 3) as well as the availability of enabling features in a mobile network and the coverage and capacity of a 2.5G or 3G mobile network all contribute to a sober decision in designing a new service. The simple choice of adopting either SMS or MMS as a delivery mechanism for content, or deciding whether or not to implement multiplayer functions in a game has a massive effect on the addressable market for a service, not to mention the complexity of design, verification and integration of such a solution. (See the case study on games developer Mforma in Chapter 6 for information on how they have tackled the complexity of games design for multiple handsets.)

It is therefore important to think carefully about the required features of a new service and its overall nature. The following basic categorization of application-to-person (A2P) services can be useful to understand the main considerations in the early stages of conceptual design. A2P refers to content- or application-based server-side services directed to individuals or groups. Application functionality can be loosely classified in the following areas:

- Server-side applications;
- Streaming media applications;
- Browser-based applications;
- Device-only applications;
- Network-enabled mobile applications.

It is also important to see this classification of A2P services in the bigger picture where they live alongside and often interact with person-to-person (P2P) services. P2P services refer to services revolving around interaction between two or more individuals and include traditional features such as voice calls, standard SMS messaging as well as more sophisticated services such as multimedia messaging and mobile e-mail. Moving forward, IMS-based content sharing services such as white-boarding and push-to-talk or image sharing services appear in the P2P domain.

To support developers to make the right choices, many telecoms vendors provide extensive information through their developer programs on the finer details of application development in the mobile environment. Sony Ericsson and Nokia for example made a very strong commitment to developers for device based development through their respective programs – the Sony Ericsson Developer World (developer.sonyericsson.com) and Nokia Forum (forum.nokia.com). Ericsson has also supported network, enterprise and total solution related aspects through the Ericsson Mobility World program (www.ericsson.com/mobilityworld) by establishing local centers in various countries to support and interact with media and content players as well as application developers.

4.2.1 Server-side applications

Server-side applications encompass many of the traditional mobile services in existence that require no specialized features in mobile devices and simply rely on **standard capabilities and usage of a mobile device**. Interactive voice recognition (IVR) services,

SMS voting, call-control, directory assistance and voicemail are all examples of services that are entirely network/server-based and generally not reliant on specialized features of mobile devices.

While the nature of these services may not seem dazzling in terms of technical innovation, the ability to address the widest possible subscriber base and the lack of expertise required to use a service such as SMS voting will ensure a steady and consistent return for services aimed at a wide audience. Importantly however, as one adopts more complex server-side functions such as MMS delivery, Parlay based call control, etc., the complexity of integration into an operator's operations, not to mention the level of commercial negotiations required, also increases.

Excluding fundamental voice based assistance services, important prerequisites for development are the knowledge of specific integration points within the operator domain in order to gain access to network-based features such as SMS/MMS short number access, call-control interfaces (e.g. ParlayX) and charging interfaces.

Case study: Yahoo – 'Start your (mobile phone) engines'
Many of the most interesting mobile applications have been made possible by the innovative use of 'standard' device features to achieve new goals. To promote their new automotive web site, Yahoo commissioned R/GA to make use of a video billboard in Times Square to allow pedestrians passing by to use their mobile phones to race each other on the 75-meter screen (Figure 4.1).

Figure 4.1 Mobile gaming on the big screen.

By dialing into a premium number, bystanders could use their phone's number pad to accelerate or decelerate in a live race against another mobile user on the big screen. Simple use of DTMF tones and a server-side application to manage incoming calls and to translate received tones into skids and slides on the racing track allowed for a promotional service that embraced all mobile owners who happened to find themselves in Times Square with a few minutes to spare!

4.2.2   Multimedia streaming applications

Since 2003, higher end mobile devices have been shipped as standard with streaming media players. A look at the current offerings by major device vendors shows that the majority of midrange and even some entry-level phones support video streaming services.

Read more about the challenges of creating strong video content in Section 4.4.

4.2.3   Browser-based applications

Modern mobile browsers support XHTML content, including formatted text, color images and embedded audio as well as style sheets, which together with the enhancements in mobile displays allow for a more colorful and flexible interaction.

It is clear, however, some 4 years after the first mobile phone-based browsers appeared, that, while some services suit the browser environment, many do not. Common services on the Internet such as news, weather or directory enquiries are well suited to a mobile browser interface. More complex and interactive services, however, are not served well by the natural limitations of a 'browse and click' interface together with the effects of download speeds, latency and content rendering delays. It is no surprise that WAP-based e-mail has never taken off; neither is the proliferation of WAP-based games very high!

On the PC side, users undoubtedly benefit from Microsoft's continued hegemony in the browser area in terms of a consistent browsing experience with service providers able to capitalize on the finer features of Internet Explorer – pop-ups excluded! On the mobile side, however, content and application developers must contend with different browser technologies from the likes of Openwave, Nokia, Teleca and Opera, all offering varying interpretations of OMA WAP standards. Developers must also handle the huge variety of display sizes, resolutions and varying color depth support which can limit design and creativity options for a developer.

DoCoMo, however, used their dominance in Japan to ensure a much easier growth of iMode-compatible services by specifying a simpler HTML-based language for their custom-made iMode devices, which ensured a much simpler proposition to content creators and application developers. On the other hand, the fact that the three operators in Japan chose different mark-up languages has made life somewhat more complicated for local developers!

Case study: the challenge of entering WAP addresses on a phone

We are today inundated by adverts, billboards and packaging that highlight web addresses. It would make sense for people seeing these adverts on the bus or train to be able to access information easily on their phones by entering the address into their phone's browser. Unfortunately, it is not so simple.

First, there is the fundamental problem that simply entering a standard web address in a phone (like www.cnn.com) often does not automatically redirect the user to the 'mobile' formatted page. Instead it simply results in the phone attempting to load the standard 'PC' HTML web page, which either fails to load or renders a full web page on a small mobile display!

In addition, accessing a mobile browser and finding the correct submenu to enter a URL is not the easiest of tasks for the uninitiated. Indeed, on many branded phones, there is not even the option to enter a URL manually since subscribers are encouraged instead to rely on a walled-garden portal.

One neat solution has been the approach by numerous companies to capitalize on the presence of cameras in the majority of today's mobile phones to allow users to simply scan the image of a code, which the device can then decipher into a URL address and automatically take the user to that location. In Japan, the use of QR codes – two-dimensional 'barcodes' in which one can embed several thousand bits of information – are very popular and supported by all of the major mobile operators for easy access to information such as links, addresses and contact details (Figure 4.2). These codes are resilient to dirt and are even decryptable when part of the symbol is damaged. They also function regardless of the orientation of the image as captured by the camera. Such codes are very useful in offering users quick access to online content via a single camera shot.

Figure 4.2 A QR code rendition of 'Hello World!!' Note the three cubes in the corners that provide orientation detection.

Browser-based applications are coming back strongly after the initial struggles and today work with most handsets in the market. Via browsers, you have a wide reach only surpassed by SMS access. The maturity of this area is, however, still far behind that of the PC-based Internet world that has evolved over the last 10 years. As an example, the design process in the mobile space is still often connected to an understanding of the underlying technology, while for the fixed Internet it is created predominantly by designers – focused on content design rather than technology.

4.2.4   Device-based applications

Device-based applications refer to those services that are installable on a mobile device. It is important to differentiate device-only applications that do not connect to the Internet or interact with server-side components. Devices that allow for such applications require an open application platform such as Symbian, Windows Mobile, Sun's J2ME or Qualcomm's BREW (Binary Runtime Environment for Wireless).

The ability to deploy a custom-developed application on a mobile device enables one to truly capitalize on the display, audio storage and processing capabilities of the mobile device and opens up a wealth of possibilities for new interactive services. As mentioned earlier, the gaming industry has largely overlooked WAP- and SMS-based games and embraced J2ME. It is important to note that, as J2ME has evolved and left the original sandbox security model (where applications could not access or utilize many device functions), it is now possible to create applications that more easily capitalize on device features such as messaging, camera functions, audio and media playback.

Case study: Mobile Scope – Moorhen Camera X
Controlling games using a mobile phone keypad has always been tricky, so an innovative approach was used for the sequel to Mobile Scope's Moorhen hunting game. Instead of relying on the keypad to aim your shotgun, movements of the phone up, down, left and right are tracked through the inbuilt phone camera and translated into movement of the crosshairs across the hunting field. Innovative uses of mobile phone features bring significant differentiation and a refreshing change to the mobile applications arena.

Open device operating systems are a double edged sword – the benefits of a more colorful and interactive application must be weighed against the significant costs of mobile device development expertise and the cost of adapting a single application for multiple devices even from a single device vendor. This can present a major headache to developers and can certainly limit the functionality of an application to the 'lowest common denominator' of supported application functions across the range of target devices.

In addition, one must consider carefully the question of getting a mobile application onto a target device. As penetration of correctly configured packet data-enabled devices increases, it is a relatively straightforward process for users to download small applications over the air. One can also rely on a user having a PC from which to install the application on the device, or alternatively there are a few providers that actually provide memory cards with content and applications embedded on them. These various options must be carefully considered since a bad choice effectively closes down the practical availability of your application to the end-user. This is especially true for operator-independent service providers that have no control over the device an end-user may have or knowledge of the configuration of that device to allow for data connections to an independent portal to administer purchase and download. (See Chapter 6 for more details on these various approaches.)

4.2.5   Network-enabled mobile applications

So called 'client–server' applications represent the logical enhancement of applications installed on devices – the incorporation of communication functions that allow for the establishment of data (or voice) connections for the exchange of data between the application client and a server. The limiting factor has been the relatively slow data connections available (GPRS, GSM dial-up) as well as the limitations imposed by early operating systems that did not allow for flexible data session handling. As devices come equipped with faster data capabilities, and as J2ME, Symbian and Windows operating systems allow for more sophisticated connections (e.g. socketed TCP and UDP), we now see many applications exploiting IP connectivity to create truly compelling services.

Case study: evolution of mobile gaming
As network connectivity improves, so do the possibilities for more advanced interactivity in previously stand-alone applications. In its most rudimentary form, we see the capability for gamers to upload high scores and create competitive communities. As latency and throughput performance improves, turn-based multiplayer games can be enhanced into more real-time multiplayer sessions. A key component of multiplayer gaming is the ability to fully interact with other people – everything from meeting people to screaming abuse at them during game play. It is critical to enable this to the maximum extent possible – in this case through IMS-enabled voice chatting and presence-enabled communities to enable easier discovery of potential gaming opponents.

On the enterprise side, with full capitalization of a permanent data collection, managers can imprison their employees with a ball and chain – otherwise known as a Blackberry – to ensure longer work hours and out-of-office contact though instant push e-mail and calendar synchronization onto the mobile device.

In this leap of sophistication one must, however, consider the effects of relying on a wireless data connection for the exchange of information. Applications must be robust and thoroughly tested to ensure they can handle interruptions and lost connections as well as varying bandwidth and latency, whilst still allowing a user to make use of the features of the application without unnecessary disturbance.

Many operators additionally complicate the picture by imposing limitations on the types of data that are allowed through a data connection towards servers that are outside of the walled-garden environment. In some cases it is not possible to establish streaming data sessions as well as use nonstandard TCP/UDP port numbers. This stems from the use of firewalls and other filtering mechanisms in the operator network for strategic and security reasons as well as the fact that there are of course a limited number of IP addresses available for distribution in any one operator domain – it is not likely that the Internet can spare some 2 billion IP addresses for packet data-enabled phones to remain permanently connected with a public IP address. IP version 6 (IPv6), will add more IP addresses but will still not be widely used in handsets at the application layer for years to come. More tips on handling wireless connections are covered later in this chapter.

As seen in the gaming case, a stepwise approach is recommended when network-enabling applications in order to better handle service complexity as well as to move in parallel with device and network advances.

4.3 KEY FACTORS OF STRONG MOBILE APPLICATIONS

No matter what the type of application is, be it voicemail or mobile gaming, there is a set of common design principles, which should be applied by application designers to ensure the best possible user experience and uptake of a service.

4.3.1 Easy access

Considering the importance of understanding the consumer as covered in Chapter 2, services *must* be easily and quickly accessed – with a minimum number of clicks. Services *must* be intuitive and easy to use – they cannot require someone to read a manual before use. Services *must* anticipate the needs of a mobile user who does not have hours or even minutes to search for and retrieve the necessary information from a service.

A simple model can be applied for developers to rate their services in terms of speed and ease of use, 0, 1, 2, 3:

- *0 Manuals* – a mobile user should be able to instantly use a service with no prior learning required and absolutely no special device configuration.
- *1 button access* – a mobile user should get easy access to a service. Operator portal access is generally available to users through a single click on the device, often from a hard-coded key. It is of critical importance that a user should *not* need to first log into services, select regions, select their device, etc. – they should where possible get direct access to a service.
- *2 seconds access* – while in many cases this may be rather optimistic, it is important to understand that services or applications cannot take a long time to load – waiting 20 seconds for a WAP portal homepage to load – no matter how colorful and fancy – is not acceptable.
- *3 clicks to service* – It is imperative to minimize the number of user inputs required to execute a service. Applications must be optimized to anticipate a user's common actions. This can be something as simple as displaying the top headlines on the front page of a portal instead of obliging a user to select 'News', then 'World News', then 'Top Stories'.

This may be an overly simple model to address accessibility, but it does raise some valid aspects that are further explored in Chapter 5.

4.3.1.1 Instant gratification

Looking at the more accessible mobile services, one can see many examples of smart feature design that accommodates the nature of a mobile user who will access a service only for 10 minutes on the bus while they have nothing better to do.

Mobile games are often designed around a concept of instant gratification. Usage 'sessions' are short and it should not require a gamer an hour of continuous play to achieve an objective. In addition, strong games allow users to automatically save progress and resume games at the stage they left them between playing sessions. This in turn creates 'loyalty' to a game and ensures a user will come back for more. As we saw in the Mansion example in Chapter 1, a session can even be started on a PC and then seamlessly continued on the mobile.

Figure 4.3 How fast is fast enough?.

4.3.1.2 Personalization

One must also consider the relatively limited IO (input–output) capabilities of a mobile device. While voice recognition capabilities are not embedded in devices, and until holographically projected displays are a standard feature, one must consider the fact that inputting data using the keypad can be tiresome, and browsing long pages of information before you can access data which is relevant to you contributes to an unpleasant experience. For that reason it is imperative that applications employ *personalization* to improve the end-user experience. A simple example is remembering a user's previous data entries, which ensures a minimum of necessary input and prioritizing the display of information to that which is relevant to the user (e.g. prioritizing ringtones that match the browsing and purchase habits of a user).

While there is a limit to the processing and storage capabilities of a device-based application, the server-side components of a networking mobile application can easily handle such functionality. Anticipating the needs of a returning customer and allowing for personalization of a service are key differentiators in the attractiveness of a service.

Case study: Eniro – "Yellow Pages"
While launching and maintaining a very strong browser-based "Yellow Pages" application for searching addresses and business details, Eniro created a J2ME application to replicate

this service in a device-based application environment. Immediately, the appeal of the service became much more clear. For a start, the service is more easily visible and accessible – presented as a recognizable branded icon on the phone desktop as opposed to being tucked away as a bookmark in the WAP browser. When one starts the application one has a more intuitive user interface and, critically, the application remembers key information such as previous search results as well as prefilled forms, meaning it is not necessary to always enter your home street or city to narrow down a common search. This ensures a more efficient and speedy search.

4.3.1.3   It is still a mobile phone

Above all, one must not forget that we are designing services that act through a mobile phone – a device that still will be used (often simultaneously) for making phone calls and sending messages!

Figure 4.4 It's still just a phone.

It is absolutely critical that an application does not interfere with or preclude a user from making use of their phone to receive phone calls or to receive and reply to messages.

For this reason, it is a prime requirement of the mobile industry-driven 'Java Verified' program (www.javaverified.com) as well as the 'Symbian Signed' initiative (www.symbian.com/symbiansigned) that applications must always be able to be interrupted by incoming phone calls and SMS alerts. The application must satisfactorily pause and allow the user to communicate and then the application must also restore at its previous state. Not only does this apply for incoming calls but also for other alerts such as calendar requirements, low battery warnings as well as physical actions such as users closing clamshell devices and profile settings that enforce a mute status thereby disallowing any speaker sound (while reminding the user that sound is muted). Read more about verification aspects in Chapter 8.

Checklist: design applications for the mobile user

☐ Services must be intuitive and easy to pick up.
☐ Allow instant gratification – quick access to services and information.
☐ Implement a proper 'Pause' function – remember previous actions and allow the user to resume where they left off.

☐ Personalize applications to enhance user experience and minimize necessary data input.
☐ Do not allow your application to restrict standard phone features such as incoming phone calls and messages.

4.3.2 Design principles

In addition to anticipating the nature of mobile users there are also many common sense principles that are important in the design of any mobile service. As mentioned earlier, aspects of mobile usage are strongly associated with instant gratification and quick responses – the ability to make a phone call to someone on the other side of the world, to send an SMS and get a faster response than e-mail. All of this contributes to an environment of expectation and, given the relatively high reliability of voice calls and SMS, there are similarly high expectations on the performance and accessibility of any other mobile service.

4.3.2.1 'Where's the start button?' – Application Discovery

The first step in making use of a service is often the hardest to optimize – finding and starting an application. For example, in the case of the Yellow Pages application, there is a major difference in exposing access to such an application by having a colorful (and recognizable) application icon on the main menu of a mobile phone as opposed to obliging users to open their WAP browser, select the Bookmarks page (if it exists) and then load the bookmark or, worse still, obliging them to manually enter a URL.

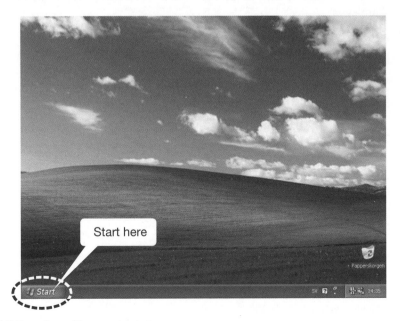

Figure 4.5 Sometimes it's easy to start.

As mentioned above, for downloadable applications, access to these services is available generally from the main 'desktop' of the phone or from within the games or applications submenus of a device. It is possible to associate graphical icons with an application, which is an ideal opportunity to capitalize on your established brand to make your service easily recognizable for the user. Some devices allow users to customize the layout of their phones and to choose shortcuts to commonly used functions. For enterprise service owners, this can be a possibility to preset devices to have relevant applications within easy reach of users. Operators also have increasing influence to make customizations to their phones and can also preset phone menu items to their portal homepages as well as prioritizing access to other popular services. This all increases the likelihood that a user will access a specific service.

In terms of optimizing access to WAP services, it does not get much better than bookmarks and access to preset links. One should ensure that a WAP-based service allows a user to bookmark the root page of the service by not embedding session-specific information in the URL of a page. Users should then be given the hint to bookmark a specific page to allow for easy access later on. Ideally, it should also be possible to bookmark a subsection of a popular service. For example, over the last 3 years, my browser's homepage has always been preset to the 'top stories' subarea of the excellent BBC News WAP service: http://news.bbc.co.uk/mobile/bbc_news/top_stories/index.wml.

4.3.2.2   Ask a user to log in and they may well just log out

The requirement for a user to log into a service is a frustrating and time-consuming process, which totally detracts from the accessibility and ease of use of a service. Studies show that each authentication challenge made to the end user leads to a successive thinning out of the user population that actually reach the service after negotiating each of the login hurdles.

Figure 4.6 Single sign-on smooths the way.

Where possible, one should capitalize on the mobile network's 'single sign-on' functionality, as described in Chapter 7. This allows for a server to render a personalized page for that user as well as automate the log-in procedure to a portal or service. Access to this feature is typically limited to service providers that are trusted or associated with the operator that owns the WAP gateway infrastructure, since this otherwise could represent a security breach in allowing the free spread of a user's phone number when they browse different WAP sites.

An easy way to enable personalization features for low-security services is to require a one-time log-in and then allow the user to bookmark the 'logged in' homepage. In such a case, the bookmark includes the user's username and password in the URL string.

4.3.2.3   Ease of use, speed of use

Once in an application, the same principles of ease and speed of use apply. Application features must be easy to find, clearly labeled and must have a consistent look and feel. This applies particularly to WAP services and portals that may comprise a variety of third-party services. A consistent interface with similar menu structures, 'home' and 'back' links, header-images, etc., insures ease of navigation.

In addition, a flat hierarchy of links in a WAP portal implies a lower number of clicks required to execute a transaction (e.g. purchasing a ringtone). The same applies for device-based applications – it must be extremely easy to find the main functions of an application without getting lost in a maze of menus and forms.

Telecoms notation should be avoided at all costs, for example, obliging a user to 'Enter your MSISDN' will undoubtedly leave users confused. Cryptic error and dialog messages are also to be avoided – 'HTTP 404 error: content could not be accessed' is not of great help to the average user. Read on in Chapter 5 on how to take this to the next step.

4.3.2.4   Sore thumbs, sore eyes – input/output

Given the relative hassle of inputting large amounts of data as well as the limited display capabilities of a mobile device, one must optimize both the methodology of data input as well as capitalize on the capabilities that exist to display information.

When inputting data, it is of enormous help to provide users with a selectable list of likely inputs instead of always obliging them to manually enter data from scratch. Prioritizing common entries (such as capital cities in the list of target localities for a search) reduces the need for text input and speeds up selections. In the same way, remembering previous selections is most helpful.

When there is no alternative but for users to enter data in text boxes, capitalize on the relatively large server-side power available to an application by allowing a user to enter partial data (such as the first few letters of a company name) and allow the server to use intelligent deduction to complete the full entry. It goes without saying that a device's T9 text input capabilities should be made use of in applications.

Additional features such as customizable soft keys, device shoulder keys and scroll wheels can be of great use in the navigation and control of an application. It is important

Figure 4.7 The results of poor ease of use.

that, if possible, the use of these keys should be largely similar to their function in normal device usage in order to minimize confusion to the end-user.

When it comes to output, we strongly recommend that applications leverage all available channels – audio and vibration are also valuable channels for conveying information. When it comes to the display, the use of the full screen and full color depth is of course a great benefit of the application environment. This relates not only to 'maximizing' the application to full-screen mode that J2ME and other open device operating systems allow, but also making use of the screen dimensions to display full-screen-size images and content instead of using 'lowest common denominator' images which fit on all devices and consequently look like tiny thumbnails on larger screen devices. Poorly ported applications that either occupy only a portion of the full screen or, worse, overlap the edge of the screen will dramatically impact the usability of an application. At the same time, cramming in too much data on the screen may not worry a 15-year-old with perfect eyesight, but if we are to address a wide audience it is important to present visual content that is readable by people who are not blessed with perfect vision.

Note that, when commanding the full screen in Java or Symbian applications, it is important that some standard device main screen information is still represented to the user. While battery life and operator network are less important, the presence and use of an active wireless connection must be conveyed since data consumption can imply data volume charges, which the user may not otherwise be aware of. As mentioned earlier, the application must also allow for the display of incoming calls and messages.

4.3.2.5 'Reticulating splines...' – user perception and user control

For the last 5 years, games developer Maxis have employed an interesting approach to game loading screens. While players of the "Sims" games wait for the game to load and render levels, they are engaged by (often) meaningless progress messages that flicker across the screen – one such classic was information that the game was 'reticulating splines'. It is hard to tell whether this is simply nonsense or in fact a genuine process.

In any case, the end user is kept 'informed' and knows that something is happening while the game loads – even if the messages are not so meaningful!

Returning to the theme of patience, it is most important that users are well aware of the status of an application or service while connecting to the network, exchanging data, during purchase authentication, etc. By going further than displaying a static hourglass and instead providing a user with *informative* feedback indicating the status of an action, one can keep an end user reassured that progress is being made. Simple things like progress bars, percentage complete indicators and useful text information all reassure users that activity is ongoing, whereas a static 'Please wait' message can all too soon be perceived as an application crash.

Figure 4.8 Avoid the uninformative hourglass.

In addition, an absolute requirement on applications that do involve lengthy waits is that the user remains totally in control of an application and can quit or postpone an activity at their leisure. Here we come back to understanding the mobile user by ensuring that activities can be paused or put off indefinitely. Simple cases such as the need to pause or cancel a 10 MB e-mail synchronization when you suddenly move over from a 3G cell to a 2G cell with a significant loss of throughput illustrate the need to empower users to be in full control.

Checklist summary

☐ Make it easy to start the application.
☐ Allow users to bookmark access to WAP applications.
☐ Do not use login screens unless absolutely necessary.
☐ Simplify application layout – no need for a manual.
☐ Use clear icons, menu items and meaningful user messages.
☐ Keep the user in control of the application.
☐ Capitalize on the full screen of the device and all possible input mechanisms.

More specific examples highlighting the importance of using end user understanding to manage the user experience are highlighted in Chapter 5.

4.3.3 Capitalizing on wireless connectivity

The benefits of mobile data connectivity need no explanation. To date, this has been mostly capitalized on only by browser-based and streaming services. As J2ME capabilities improve together with the ongoing evolution of devices with open operating systems such as Symbian, we will now see many more device-based applications that capitalize on a data connection.

Consider again the case of mobile gaming. Game developers initially capitalized on the advanced J2ME environments on mobile devices to create highly playable three-dimensional games. However, the truly innovative games are those that have introduced wireless multiplayer elements (both over Bluetooth and mobile networks). By exploiting the connection to the Internet, multiplayer services in Japan have now evolved from simple high-score sharing to extending PC-based MMORPG (massively multiplayer online role-playing games) with new mobile features for gamers to keep track of their 'fixed' games while they are away from the computer on their phones. Interestingly, when die-hard players in this genre are on vacation, they have been known to pay other gamers to play with their character in order to keep their status in the game. There we have it, instant cost savings when one can access the game on the mobile and have it with you all the time . . .

4.3.3.1 Packet data connections

Evolving from traditional GSM modem 'dial-up' connections, recent network evolutions such as GPRS, cdma2000 and WCDMA provide data connections that are fully packet switched. In the GSM circuit switched world, connections were made through modem pools accessed through a dial-up number. In the packet switched world, IP connections are made through operator-controlled access points (known as APN, access point names). These provide a connection directly to the Internet, into an operator's walled-garden portal environment or even through a virtual private network (VPN) connection to a corporate intranet. These connection parameters are set within each mobile device and they specify the access points, usernames and passwords as well as proxy and gateway addresses specific to each connection. On newer mobile devices there are separate settings that specify the data connection configurations for WAP access, streaming services access and for J2ME application-based network access. Developers can invoke these connections within an application without needing to specify them manually, but a developer must insure that these default settings are suitable for the network access that the application will require – again, local intelligence on the settings an operator in a target market uses is critical.

On 2.5G networks such as GPRS and cdma2000-1x, you are typically limited to a single simultaneous data connection, i.e. a connection to a single access point. Additionally, until dual transfer mode-capable 2G phones appear, one cannot send data and make a phone call at the same time. One can, however, send an SMS and use data or voice services simultaneously since SMS traffic is sent over independent control channels. This is an important factor of 2G devices that can affect data applications and services that require a simultaneous voice connection.

It is possible, however, on 3G networks such as WCDMA to make simultaneous data and voice connections as well as maintaining multiple independent data connections. These networks have been standardized from the ground up to accommodate several different simultaneous connections and this can allow for more richly interactive person-to-person and content-to-person services.

4.3.3.2 But are the phones configured?

Operators generally provide devices with data settings preinstalled when devices are sold as a part of a subscription package, however not always. In this case, devices may not be configured to allow for packet data connections, which means that users cannot browse WAP services, make streaming sessions, send MMS messages or allow applications or Bluetooth/Infrared connected devices to establish an IP connection.

Example: True tone request from a nonconfigured phone

A user may send a premium SMS to request an MP3 ringtone. Typically, a WAP push message would then be sent out to allow the user to download the ringtone. If the user has not got the correct data settings they will not be able to download the content. In this case, the link should time out after some 30 min and the user should not be charged for the purchase. Ideally, an SMS message should then be sent to the user to give them a hint on how to configure their phone and to reassure them that the purchase was not charged.

It is an extremely frustrating aspect of mobile data networks that operator-specific settings are required for each new data technology (WAP, MMS, streaming, etc.) and that many phones, especially older ones, do not have these settings installed by default. Many operators are now launching automatic device configuration capabilities that will automatically push out default settings for new phone users. Figure 4.9 illustrates the actual improvements that a Southern European operator recorded after implementing such a system. Details on how these solutions can be implemented are to be found in Chapter 7.

Those providers that are independent of a specific operator must provide references to the correct settings that are needed to access a service or incorporate some form of device configuration.

Figure 4.9 Massive improvements with automatic device configuration. Source: Ericsson.

4.3.3.3 Data connection limitations

When using mobile connections to the Internet, access points are generally unregulated with regards to TCP/UDP port usage and the only throughput limitation is the natural limits of the wireless connection. Data traffic is generally billed by data volume or at a flat monthly rate.

There are some cases, however, where phones are configured for access directly into an operator's walled-garden portal – an IP domain that can give rise to limitations in wider

Internet connectivity. In some cases, network address translation (NAT) gateways and port filtering can affect the accessibility of external streaming services and access to other non-HTTP (port 80) based servers, thus limiting the connectivity of device-based applications.

Application developers and independent service providers should anticipate this potential complication and research the nature of each local operator's standard access points to ensure that data applications will work in these environments.

It is not feasible to expect an end-user to change phone settings in order to get an application to work. Only in the case of a customized enterprise or vertical application should the use of customized APN and data settings be considered and if these applications involve a large user base it is imperative to consider automatic over-the-air distribution (or preconfiguration) of such settings. Obliging uninitiated users to change their data settings is a recipe for disaster.

4.3.3.4   Data access when roaming overseas

International data roaming access is steadily increasing in coverage. Larger global operators and consortiums now have strong international data roaming coverage; however in many smaller countries it is still not possible to establish a data connection when roaming. Roaming data charges and tariffs remain prohibitively expensive.

Roaming data subscribers who attach to an overseas operator will have automatic access to their home APN connections – their data connections will thus function as if

Figure 4.10 Support for roaming is key.

the phone were attached in their home network. Note, this is not always the case for users of premium SMS/MMS numbers. When roaming overseas, you may not be able to use your regular SMS or MMS short numbers since these pools of premium numbers are usually domestic, independently allocated and controlled by local operators unless customized solutions have been put in place to allow for overseas use of a shortcode.

4.3.4   Wireless design considerations

Developing applications for a wireless environment introduces a number of new considerations given the restricted throughput capacity and relative volatility of data connections in a wireless environment.

4.3.4.1   Throughput

The first obvious limitation is that throughput capacity, while significantly better in 3G networks, renders many standard Internet applications unusable. Generally speaking, upload speeds are lower than download speeds since the battery power required to upload data (i.e. to make radio transmissions) is significantly higher than that required to receive data.

The impact of this is already seen in MMS, where proud owners of multi-mega-pixel phone cameras find that it can take several minutes or more to send a full-sized picture over a GPRS network. Likewise, sending large e-mail attachments can be a painful experience. Typically, in 2.5G networks such as GPRS, the upload speed capacity is between 10 and 20 kbps, while download speeds can reach 40–50 kbps. For this reason, designers should consider the overall requirements of their applications to send data. In 3G networks, there is relief on the way with immediate upload speeds of at least 64 kbps being reached with higher speeds possible for data cards and networks that support such bearers. Further ahead, mechanisms such as HSUPA (high speed uplink packet access) in WCDMA networks introduce dedicated features that further boost uplink packet throughput.

IMS-based services will certainly push the requirements for higher upload speeds since these systems expose greater possibilities for high-speed interactive P2P and A2P multimedia sessions. See Chapter 6 for more information on realistic network conditions and the penetration of devices with support for various network technologies.

4.3.4.2   Latency

Latency is well known in the PC multiplayer gaming world where game servers are rated by their ping response times. Latency refers to the delay to send packets of data to a server and back. In wireless networks there are error correction mechanisms used to compensate for data loss on the wireless link between the device and radio base station. For standard wireless data connections, these mechanisms are configured to ensure that there is no data loss on the wireless link and, for this reason, automatic retransmissions are made between the device and base station (independent of the application or TCP/IP stack) to correct data packets. These retransmissions lead to a relatively long delay in the transmission of packets over a wireless link, leading to a high latency.

In 2.5G networks, latency can be on average about 450 ms, with optimizations in new 3G networks expected to bring that gradually below 100 ms in the future.

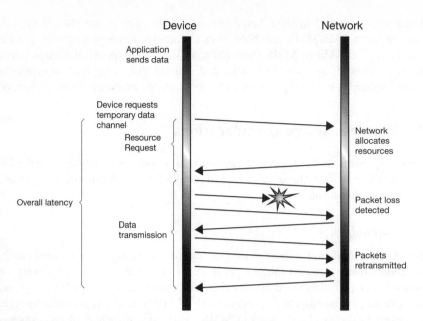

Figure 4.11 Resource requests and error correction latency diagram.

The reason for this performance is due to the additional need for wireless error correction, but also because of the fact that a mobile radio network is a shared medium where packet data channels are generally only allocated on the fly when needed. In GSM circuit switched networks, a permanent circuit is established for both voice and data connections, resulting in the fact that calls are generally billed for the time of the connection, since network resources are dedicated to that user. In packet data networks, resources are allocated as needed (hence the trend to charge based on data volumes), so even if a phone has an active data connection it must still request resources from the network to upload or download data when needed. This resource request process also takes time and creates a 'one-off' latency hit at the start of a packet exchange. For the duration of that exchange, the data channel is dedicated, but as soon as the data exchange stops, a timer is started and once that expires (typically after several seconds), a new data channel resource 'request' must take place (Figure 4.11).

Latency is an extremely important aspect of wireless performance that is not linked to the throughput of a data connection and affects large and small packets alike. Latency is particularly significant for applications that send a lot of small data packets very frequently (e.g. instant messaging applications) or for services that use 'chatty' communication protocols (such as HTTP).

Despite optimization in network technologies such as HSDPA, latency still remains a factor to be anticipated, especially in the short term.

4.3.4.3 Mobile conditions change!

Unfortunately, it is a fact of life that mobile users move around! Mobile users will thus experience varying conditions as they move nearer to and further from their attached base station(s) and of course they may lose coverage completely in tunnels or other areas of poor coverage.

Early users of 3G networks experienced relatively poor coverage due to the lack of widespread coverage. This was felt especially indoors due to the fact that the shorter wavelengths of higher radio frequencies (such as WCDMA) do not penetrate buildings so well.

In addition, the load on a wireless network can be highly volatile. In most networks, operators prioritize voice traffic over data (for revenue purposes and due to quality expectations of voice users). This also affects the capacity and performance of a wireless link, which developers must anticipate. As data service usage grows, however, we see that more and more operators dedicate capacity to data users and this results in a significantly better performance; these are the simplest manifestations of ensuring a basic quality of service. In fact, dedicating capacity for data is a prerequisite for success.

4.3.4.4   Quality of service

Wireless packet data networks have been designed with these wireless limitations in mind and are specified to support quality of service (QoS) profiles that 'guarantee' throughput and latency parameters. In practice, when establishing a wireless data connection, it is possible to request a QoS classification and the network will return with an acknowledgement based on the permissions of the user to access such a quality level as well as based on the current load of the network. These requests can be made either by an application running on a mobile device (as a part of the request to establish a data connection) or from the network side, when a data session starts and the server instructs the network to maintain a quality level for that premium service.

Customized radio bearer profiles have been used as the foundation in WCDMA networks for creating voice and best effort data channels. In more sophisticated cases they are used to provide high bit rate, low latency and low error-correction bearers for multimedia streaming services.

However, these capabilities are not thoroughly released to application developers at this time. Few devices provide APIs that allow one to specify QoS requirements and it is not clear when operators will widely release QoS capabilities for internal use or towards independent service providers. This could in the future be a source of 'premium' revenues for operators by giving best-in-class performance to high-paying customers; however, the business proposition is unclear – should I pay for a high QoS when the network is under-utilized?

Practically speaking, QoS is a difficult aspect to manage. If a user requests a 128 kbps video stream at peak periods and the QoS cannot be guaranteed due to load on the network, what will the result of the request be? Will the user be told that they cannot access the stream at this time even though they theoretically could access the stream, or will they be given 'best effort' for a discount? While these systems are not yet mature, it is all too likely that the simplest solution will prevail – capacity over-provisioning and best effort access are the simplest ways to ensure the best possible experience and monitoring of the performance of a delivered service should ensure that users are not charged for poor performance. As we will see in Chapter 10, another key aspect is to be aware of what the performance is in order to know what to promise the users.

4.3.4.5 Designing applications that anticipate wireless conditions

The simplest advice to give here is to ensure, when designing a mobile application that interacts wirelessly, that wireless functions are decoupled from the other operations of the application – on both the device and server sides. This means that a user should optimally be able to freely navigate and use an application irrespective of the quality or availability of a network connection. Obviously, when a connection is required to download content or complete a purchase, the function of the application depends on its ability to connect, but a user should never be prevented from accessing other independent application functions simply because a network connection is not available.

Case study: MMS Outbox
A simple example of this is how MMS transmission is handled on mobile devices. In the past, a user was obliged to wait while a message was sent (similar to SMS). However, given the time it can take to send a large MMS, most new phones process the sending of the MMS message in the background while allowing the user to carry on using the phone for other activities after they have hit 'send'. If a message fails to be sent it is simply cached in the message outbox for a later attempt without disturbing or obstructing the user. In effect, the user interface of the phone is decoupled from the network-dependant activities of the application, and this approach should be similarly applied in mobile application design.

Where appropriate, application developers should incorporate network functions with a 'store and forward' mentality. Network-related activities should be queued in the background and executed independently when a network connection is available. Where possible, users should not be obliged to wait for minutes while a network transaction is attempted. Simple examples of this are:

- Allowing for e-mail synchronization to be made in the background while allowing the user to compose new messages and browse existing inbox items.
- Downloading updated news content in the background while a user reads the headline page or another article.
- Allowing a user to make local updates and changes to database records while a connection to a global database is unavailable. When a connection is possible, these batched updates should then be compressed and sent.

In addition, one must ensure that applications are thoroughly optimized to only send the absolutely necessary data over the wireless link. Superfluous or duplicate information should be removed and any data that is sent should ideally be compressed to minimize throughput delays and to reduce the risk of lost data. As mentioned earlier, one must also anticipate that network conditions are subject to change. This has many causes:

- Geographical movement of a mobile user (car, trains, walking) results in handovers between cells, causing short-term interruptions and a change in cell load conditions.

- Rapidly changing load on a cell from another user's voice and data traffic will affect wireless conditions, especially throughput.
- End user actions such as incoming or outgoing phone calls can pause data connections (e.g. GPRS) or degrade their performance for the duration of the call.

Applications must be able to handle these sometimes abrupt changes in network conditions both from a server and device perspective and strong applications will adapt appropriately and inform the user of changed conditions that could significantly affect the performance of the service.

In the case of a drastic reduction in available bandwidth, possible adaptations can be something as simple as downgrading the bit rate of a video stream to match the new conditions. Another example can be ensuring that only header information of messages during e-mail synchronization is completed initially to allow a user to browse new mail headlines before a more complete sync is done of full message contents.

4.3.4.6 How to detect the quality of a connection

Unfortunately, there are no standard APIs or methodologies for developers to use to instantly diagnose network conditions – there is no way to objectively query what type of network a user is attached to (e.g. 2G or 3G) nor evaluate the load and performance of the link.

Some operator walled-garden portal technologies can diagnose what type of radio network a user is attached to from the incoming IP address subset that the user queries the portal from, but otherwise it is up to the application itself to self diagnose the connection – potentially by measuring the speed of a connection by analyzing packet flow. As device API libraries improve and as the introduction of QoS features becomes mature, it will be possible to request the establishment of a data bearer with specific QoS features such as high bandwidth or low latency and the acknowledgement from the network will provide feedback on the quality of the bearer provided. Until this is mature, however, a manual check is required. Alternatively, developers must make a conservative estimate of generally available conditions and tailor their application appropriately. In any case, good research and target market testing of a service is needed to guage whether it will perform as it is expected to.

4.4 CREATING MOBILE MEDIA SERVICES

For the end-user, the introduction of many new multimedia capabilities in the mobile space has unfortunately resulted in a relatively poor mobile experience as device vendors have only gradually implemented support for the different image, audio and video formats with each vendor maintaining their own roadmaps for support and different classes of devices implementing various subsets of features. Often, the lack of support for these formats on traditional desktop computer systems has created frustrating incidents when people try to send audio or video clips from their mobile to a friend's PC e-mail address only to have the recipient be unable to play back the attached .AMR audio or .3GP video content on their PC since Windows Media Player does not recognize such content.

In the early days, media industry content creation tools took time to offer mobile media plug-ins and thus mobile developers were instead left to deal with a variety of content creation tools provided as a part of device vendor SDKs. Some tools were up to the challenge, but many were sadly lacking. All were naturally proprietary and addressed only that vendor's subset of phones.

4.4.1   A new age of mobile content creation

Standards such as WAP have evolved to be more in line with the rest of the content world and recognized tools exist such as Adobe's GoLive and Macromedia's Dreamweaver that support both web- and WAP/iMode-based content creation using the harmonized xHTML content standards. Content creation in the mobile arena is a far easier prospect these days using these tools created by companies whose 'bread and butter' is the business of delivering tools that support content creation and that allow easier development of unified content for both the mobile and fixed space.

Perhaps an overly simple, yet symbolic, sign of the times is the fact that, when designing xHTML-based WAP 2 content, it is possible to preview and try out the look, feel and flow of pages within a standard PC web browser like Internet Explorer. While it is not possible to assess the performance of the content on the limited screen size of a phone (despite resizing the Internet Explorer window much smaller), it is much easier to prototype portal flows and ease content development than it was some years ago.

Video and audio content creation is also settling with wider support for 3GPP standards in the hand-held device industry and portable players and devices such as Sony's Playstation Portable supporting equivalent MPEG4 standards as video file formats. However, as we will see later, with the harsh requirements for limiting content size for mobile devices, this often results in a significant reduction in quality.

4.4.2   Content creation and control

Looking into the typical process of creating content for example a TV series, we arrive at a (simplified) view shown in Figure 4.12.

A company with existing content that moves into the mobile channel often starts with the existing content in order to convert it to mobile. Since this only works to a certain extent and does not provide an optimal user experience, we see more and more how the

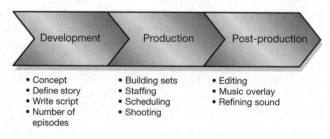

Figure 4.12 Holistic view of creation process for a TV series.

mobile channel gets integrated into the regular content creation process (see Figure 4.12). When integrating the mobile production into this process, the usual approach is to start from the right in this figure (post-production) by training the editors to cut specific mobile clips while producing the regular shows.

As a second step, you can then move from post-production into the production part to adapt it for the specific needs of mobiles. This can include shooting specific angles or more close-ups that would look better on a small screen. Shooting a dialog with a snowing background looks good on a widescreen TV but only blurs the image when put on a mobile. Another part of the production phase is how the set is built and light is arranged. This can be used to ensure that proper contrast between colors is achieved as well as minimizing the number of objects that could disturb the experience.

Finally, the left-most step in Figure 4.12, development, can be adapted to fit the mobile channel needs. At this stage, even the script is written to leverage the capabilities and needs of the mobile channel. One approach is to build mobile episodes that serve as optional extensions to the regular show, but deepening the value for the real fans. In summary, the show can be conceived with the mobile in mind.

Case study: Chum Television creating two music magazine shows for their channel MuchMusic
Chum Television has come a long way in mobilizing its content and creating interactive experiences. Chum Television, with their channel MuchMusic, has started cutting 90 second mobile shows at the same time as doing a full hour-long television magazine show. For this 90 second clip the camera angles and lighting are chosen during editing to fit the mobile channel, e.g. including more close-ups of the people involved. Additionally, the mobile version often has a different editorial approach to it. A 60 minute show featuring an interview with Britney Spears for the TV could at the same be edited into a 90 second special where Britney talks about a specific theme, for example what it is like to be called a sex symbol.

Provided the budget allows for it, a greater investment in the early phases of content creation to take mobile aspects into account will always lead to significantly enhanced material for use in mobile channels.

4.4.3 Appropriate mobile content

Given the limitations of mobile devices, the needs are clear for considering carefully the use of content in mobile channels, and this means not simply reformatting and directly reusing image, audio and video content from the mainstream media world. Disney, for example, took this concept one step further when they launched Disney DIMO on i-Mode, with characters created exclusively for the mobile scene.

4.4.3.1 Image content
The perceived quality of images used as phone desktops, in MMS alerts or as a part of news content is dependent on the cornerstones of good photography – good lighting, a quality camera and of course a great photographer.

In a mobile context, given the limited screen size of mobile handsets, one must also have a strong focus on the main subject of the photo. When considering the variety of screen sizes, dimensions and screen proportions it is, however, important to take a photo with ample space around the main subject to allow for cropping of the image to allow for this variety of target displays.

More time-consuming than the original creation of image content is the maintenance of image-related services. One needs to continually adapt image content to support the many new devices that are launched on the market while of course maintaining support for all older devices.

This implies that original material should always be stored in the uncompressed raw format so that such master copies can be easily manipulated into customized images suitable for each and every target device. Display sizes and resolution are continually improving so it is insurance of a sort to maintain a database of images in their original high-resolution, uncompressed state.

Looking at the many image formats that mobile devices support, it is accepted that JPEG images provide the best quality with very few compression effects. GIF images will often render faster on a mobile device; however these files are limited to 256 colors, which may not be sufficient for certain images. Animated GIFs are, however, becoming more and more popular in the mobile scene, although one must be careful to consider the overall file size of such content.

Detailed or varied backgrounds should be avoided when creating image content. Such features can make the image harder to view on smaller screens and also lead to a less effective compression of the image, with the quality of the main subject of the image possibly suffering.

4.4.3.2 Audio content

As the mobile world embraces the potential of music services and devices appear with storage and playback capabilities that equal traditional digital music players, the challenge remains to provide audio content that is of high enough audio quality while remaining a reasonable size to download.

Excluding synthetic audio formats such as MIDI (Musical Instrument Digital Interface), there are a number of 'natural' audio formats used both as a part of streaming content as well as downloadable file formats. AMR (adaptive multirate codec) and AMR-WB (wideband) formats were originally created for enhanced sound quality for speech signals. These formats both provide mono channels only. It is only with AMR-WB+, introduced in 3GPP release 6 in 2004, that stereo support is added as well as bit rates up to 48 kbps. However for quality music playback, MPEG-4 AAC (Advanced Audio Coding), and especially AAC+ and enhanced AAC+ (EAAC+), encoding at rates of 64 and 128 kbps, as well as Microsoft's WMA (Windows Media Audio) formats are the ideal formats providing both high perceived quality and extremely small file sizes – full music tracks as small as 800 kB.

4.4.3.3 Video content

When considering the nature of compression engines there are several tactics for video content creation that can greatly improve the end video product. Ideally, you need to keep the camera still when recording – if possible, take one shot at one angle, then another shot at another angle. Do not use fades between scenes and, again, use simple backgrounds. All of these approaches reduce the demand on the video compression engine in the encoding phase and will result in a higher quality end product.

Content creators must consider the ability of a mobile device and the related compression engines to effectively display fast-moving content (cricket balls, base balls, golf balls, etc.) as well as rapidly changing content such as movie trailers. Tickers, subtitles and small text content, while appropriate for a 32-inch screen, are unreadable on a mobile display and destroy the efficiency of a compressible video stream. Optimally, overlay content should not be included in a mobile video channel but instead conveyed separately and displayed independently of the video.

Case study: Interactive Mobile TV

There are clear synergies in the great potential in mobile TV services together with the existing success of SMS voting and greeting services that have become a *de-facto* feature of many TV programs. Ericsson, together with partners conVISUAL and Communology, showcased a mobile interactive TV solution where users watching TV on their mobile phones could easily send greetings and votes as well as purchase related content with a single button click (Figure 4.13).

Such interactivity was enabled by the use of a customized mobile TV application that when launched, would establish a streaming TV session as well as connecting to a synchronized information feed that would provide interactivity prompts based on the current TV program contents.

The solution makes it possible for viewers to interact with a show they are watching on their mobile device in a whole new way, creating a much richer TV experience with the help of the mobile channel. Apart from simultaneously watching and interacting with TV on a mobile device, the solution also allowed the device to be used as an interactive

Figure 4.13 Interactive mobile TV.

remote control when sitting in front of the TV at home. TV networks can profit from content fees, additional advertising revenues and paid interactions such as voting, greeting and shopping.

There are a variety of 3GPP-endorsed video codecs with support for new variants appearing in new devices regularly. For that reason, it is not worthwhile pointing out specific aspects with regard to encoding format other than the obvious need to follow the standard and closely verify device vendor specifications for true support.

When it comes to video downloads one must carefully consider file size with anything between 100 and 300 kb being reasonable for 2.5G download rates. This translates to between 10 and 90 s of video. A key limitation in many networks is that many WAP gateways do not support delivery of files larger than 300 kB. Furthermore it is often very hard to detect whether a device is configured to access the network directly (WAP 2 devices can access services directly over HTTP) or via a WAP gateway. For this reason, one is sometimes obliged to blanket limit the size of video content to ensure there are no unnecessary download failures.

As throughput rates increase with 3G rollout and download limitations are sorted out, the potential arises for longer video clips. However, consider also the fact that device displays are becoming larger and their resolutions are increasing – this means more video data is needed in order to 'fill' the screen, implying that file sizes will naturally increase over time in any case.

Streaming is of course an alternative approach for the delivery of video content. In this case one must carefully select the appropriate encoding bit rates that will result in a reliable performance based on a conservative estimate of available network bandwidth.

In terms of file size limitations for downloads and bit rate limitation for streaming, one can achieve trade-offs in a more aggressive compression of the video clip, lower rates of encoding and fewer frames or alternatively one can degrade the audio quality of the clip using a simpler AMR codec. Such a choice is dependent heavily on the subject matter and the relative importance of the video and audio components.

4.4.4   Content in the larger picture

When creating a portfolio of content it is necessary to consider the requirements for providing content for use in promotional material such as content catalogs and databases as well as external marketing, such as printed ads in magazines, web page placement and even TV commercials. It is also important to anticipate the use of preview content that can be used as a teaser within a mobile or web portal to better stimulate sales. Again, the benefit of retaining content in its original format serves itself here where it is of use for promotion of the related services

4.5   SUMMARY

When planning the design of a service, make sure you do not get carried away with all of the possible features and functions that you could throw in. We will see later on when it comes to deployment how one benefits strongly from a focused service that is as simple to integrate as it is simple to use. Choosing too many technical features may also limit the addressable market for your service

Consider carefully important factors such as wireless optimization as well as the many finer details of creating content for the mobile channel. Taking the cheap and easy approach by simply repackaging existing TV or web material will damage the customer perception and overall robustness of your service.

5

Managing the Customer Experience

5.1 WHAT THE CUSTOMER EXPERIENCES

Jack switches on the device and within seconds the screen lights up. He sees a newscaster reading the latest news. The picture switches to a video clip from a remote region of the world illustrating the story. Jack pushes a button and a menu of up to 100 different channels appears. He scrolls down and sees that in 10 min there is a movie starting on channel 23. Jack skims some of the other channels just to see what else is on but finally punches in '23' and settles down to enjoy the film. Jack is watching TV.

Jo switches on the device and instantly a moving picture appears on the screen. Jo positions the device to frame her 7-year old daughter and pushes a button. Silently the device scans the scene with an infrared beam, measuring the distance to the child, and adjusts the focus. It takes several simultaneous measurements of the light levels at different parts of the image and adjusts the aperture/shutter-speed ratio for the optimum exposure. One-hundred milliseconds after Jo pushes the button the image on the screen freezes and she turns the device towards her daughter, who laughs. Jo is taking a photo using a digital camera.

Nelson switches on the device and a menu appears. He runs his finger over the circular front panel and scrolls through his music collection, sorted by artist and song title. He highlights a song and presses 'Play'. He places an earpiece in each ear and the noise of the bus is replaced by the sounds of his favorite band. Nelson is listening to music on his portable MP3 player.

These three scenarios describe everyday situations that most of us will have experienced. They have several things in common:

- They each provide an end experience highly valued by the customer – watching a movie, seeing a child's photograph, listening to a song.
- They are all modern versions of well-established 'applications' and build on established behavior – people have watched TV, taken photos and listened to music for decades.

Mobile Media and Applications – From Concept to Cash: Successful Service Creation and Launch   Christoffer Andersson, Daniel Freeman, Ian James, Andy Johnston, Staffan Ljung
© 2006 John Wiley & Sons, Ltd

- They all use complex technology and yet the customer experience is separated from the technology – the interaction is kept extremely simple, involving a push of a few buttons or the simple scrolling of a menu.
- All three scenarios are today also possible using the latest mobile phones.

The mobile phone itself provides its own similar scenario. When I switch it on the screen lights up. I type in the phone number of a friend and press 'Call'. I wait for my friend to answer and start chatting. Again, complex technology but very simple to use and built on established behavior, both in terms of the end experience – the conversation with my friend – and in the simplicity of entering the phone number and dialing.

In all these cases the device, whether television, camera, MP3 player or mobile phone, was developed with a specific application in mind and is optimized for that application. When, however, we start to combine all of these functions into a single device, we inevitably compromise the end experience (watching TV on a phone, for example, will never be a substitute for the real thing) and we risk exposing the complexity of the device to the person using it – the customer.

This chapter is called *managing* the customer experience. The customer's experience is *our* business, *our* responsibility. We have to work from two directions: managing both the positive experiences and the negative; the drivers and the barriers to use. We must make the drivers as attractive as possible and take steps to eliminate the barriers. Remember, it does not matter how many drivers there are – in fact the more the better – but just one barrier will prevent our customers from using our services.

5.2 MAXIMIZING THE EXPERIENCE

The customer experience is everything. Jack switched on his TV to watch a movie; Jo bought a digital camera to take photographs; Nelson uses his MP3 player to listen to music; I switch my phone on to make and receive phone calls. The technology used in each case is simply a means to the end, and it is the *end* that we as customers value.

Even though these devices are easy to use in their normal use, there is often some complexity that we have to deal with in order to get the full use out of them: a TV has initially to be connected up and tuned in to receive the available TV channels, not forgetting the need for the programs to be produced and broadcast in the first place; the photos in a digital camera must be transferred to a PC or portal printer in order to display them on a large screen or to make paper copies; music must first be transferred from CD or downloaded from the Internet onto a PC before it can be loaded into the MP3 player. In each case the value of the end experience makes it worth the effort required to master the technology.

As a basic telephone the mobile phone is in fact simpler to use than any of the above devices. Once the preconfigured SIM card is inserted and the phone switched on customers can start making calls and sending text messages with little or no additional user-intervention required. The challenges begin when we start to widen the uses of the mobile phone.

Complexity inevitably increases considerably as we add more and more functions to the phone and as we expand the range of content and applications with which the phone must interact. The biggest challenge is to ensure that customers continue to value the experience that this new functionality offers sufficiently to overcome the added complexity. Value can be measured in many ways; let us consider some.

5.2.1   A rich experience from basic content

We should not confuse the richness of content with the richness of the customer experience. It is a common mistake to focus only on technology's cutting edge to produce our applications and forget that humbler resources are in many cases more than adequate, and in fact preferable, for many applications.

The obvious example, as we learnt in Chapter 2, is SMS. A timely text message has huge value for people in countless situations, from changing the time of a meeting, to voting to remove the most obnoxious Big Brother contestant. It is the same with basic information services: the value is in the information itself not the format in which it is presented. One of the most useful WAP sites is BBC News, which through simple text pages allows users to keep constantly up to date with world events. What is more, text can be quickly downloaded so the customer gets the information fast.

5.2.2   The quality of content and applications

5.2.2.1   The creator

Producing content and applications is a creative process. Polyphonic ringtones, for example, must be composed using a range of different instruments. Just as with actual music, the ability of the person composing the tone determines the quality of the result. It is similar with other forms of hand-produced content such as graphic images, cartoons and games; the talents of the designers directly affect the quality, and therefore the value, of the product.

5.2.2.2   The handset

Different handsets have different capabilities. The value of a content item or application can be greatly affected by the characteristics of the handset on which it resides. For instance, most modern handsets can play four-channel polyphonic ringtones (i.e. a ringtone composed with four different instruments), but some handsets can accommodate ringtones with 16, 24 or even 40 channels. Similarly screen size, image resolution and color capabilities vary widely between handsets. To maximize the value of content and applications, they should use the full capabilities of the handset in which they reside. Content providers must create content in a range of formats, and the operator or service provider supply the item best suited to each handset.

5.2.2.3   The network

The mobile network itself affects the customer experience of bandwidth-hungry content such as videos and music.

The value of a mobile video experience is very subjective and is a balance between the quality of the content and the time it takes to download. Video image quality is determined by the bit-rate and frame-rate at which the clip is encoded. Video encoders process only the changes from one frame to the next, so with fairly static images such as a person reading the news the image can be updated more regularly than images with a lot of movement, where the increased time between frame updates produces a jerky image. A higher encoding rate produces high-quality images, but the resulting file may become so large it could take several minutes to download a clip over a network such as GPRS. Shortening the duration of the clip reduces the file size but then the value of the clip is diminished.

Faster networks, such as EDGE or 3G mean quicker downloads, allowing files to be larger and so longer, and/or higher quality video can be supplied. CD quality music tracks result in a file size of 1 mb per minute of music, so full music track downloads are only practical with these higher speed networks.

When it comes to audio-video streaming, the quality of the audio track is in fact more important than the video, from the customer experience perspective. This is because most clips have an audio commentary – if the images break one can still listen to the commentary, but if the audio stops we lose the meaning of the images.

5.2.3   The power of brands

When we talk about something's value we often mean its *perceived* value. Nothing illustrates this better than the concept of brand. People will often pay more for a product with a recognized brand than a similar and cheaper but nonbranded product. An example from the mobile world illustrates this nicely.

Case study: Virgin and One-2-One
In 1999 Virgin in the UK became a mobile virtual network operator (MVNO) in collaboration with then mobile operator One-2-One (now T-Mobile). Virgin offered mobile services, branded as Virgin, over One-2-One's network. This worked well for both companies: Virgin could offer mobile services without requiring a mobile license or network; One-2-One gained more traffic on its network.

A subsequent customer survey measuring people's opinions of UK mobile providers revealed that customers consistently rated the quality of Virgin's 'network' higher than that of One-2-One's, even though they were one and the same thing. It was the strength of the Virgin brand, perhaps influenced by other customer-related services, which created a greater perceived value.

In the same way, branded mobile content is perceived as valuable. For example, offering news content from BBC, CNN and the Financial Times not only ensures that the content will be of high quality, but that customers will perceive it as being so, more than if an operator offers the same news stories under its own brand. This effect will be multiplied if, as described in Chapter 4, an icon representing the brand appears on the phone's menu.

The possibilities are endless: Orange UK offered ringtones, images and games in conjunction with the release of the final Star Wars movie. Some of the biggest selling mobile games are mobile versions of established PC and game-console titles. Popular TV cartoon characters make ideal mobile content, both for their brand power and because they target the key youth segment.

5.2.4   Time is value

The value of content or an application is often determined by its timeliness – its relevance *now*. Examples are financial triggers on a specified stock price, an alert of a goal scored or a list of what's on TV tonight. All these have value for a limited period, after which their usefulness or relevance rapidly drops to zero.

Another aspect of time is the freshness of the content. We have already mentioned the BBC's WAP site. This is updated on a minute-by-minute basis so users know that the information displayed is the absolute latest data. The customer experience is one of being constantly up to date and it means that people are more likely to visit the site on a regular basis.

This principle applies even to generic content such as ringtones and wallpapers. If the content on offer remains static, such that every time customers view a portal site they see the same items as before, they are unlikely to return very often. Think how you would feel if the same programs were shown on TV every night.

5.2.5   The small screen experience

We are often required to adapt content to suit the limitations of the mobile phone. Sometimes this is easily done by shrinking images or compressing video clips, but sometimes we have to understand the essence of a piece of content in order to maximize the customer experience. A simple example is a ringtone, which generally gets no more than 10 s to do its thing before the phone gets answered. The person who composes the ringtone must select the 10 s portion of a song that makes it instantly recognizable. Often this is the chorus – Abba's 'Dancing Queen' springs to mind for some reason; sometimes, as with Coldplay's 'Clocks', it is the opening few bars. On odd occasions there is nothing the composer can use – some rap music for instance has no distinct melody (some might say no rap music has a distinct melody), which makes it impossible to produce a recognizable ringtone.

UK operator 3 discovered that our experience of content on the mobile was very different from other media.

Case study: Premier League live video
UK operator 3 bought the mobile rights to broadcast live clips of UK Premier League soccer matches as the spearhead of the company's 3G launch. The initial plan was to transmit TV feeds directly to the mobile, but in trials it was found that the experience of watching mobile soccer just wasn't as exciting as they thought. Further research revealed that customers preferred close-up shots of the players, or the expression on the manager's face after a goal was scored, to the wide-screen shots typical of television.

The mobile phone provides a very personal experience and content designed for a larger format does not necessarily translate well to the mobile phone without modification, particularly if, like soccer, the content is normally experienced in a social situation.

In fact, despite adapting the content for the small screen, few people actually used the live service. People who really wanted to see the match watched it on TV or went to the game itself. 3 changed the service to provide end-of-match video highlights, which were far more successful. This reiterates the need for timely content, and that sometimes it can be too early!

5.2.6 Targeted customers

In Chapter 2 we described the need to understand our customers and we learnt that different customer segments have different interests and needs. So the obvious way of maximizing the customer experience is offer content and applications tailored to each customer segment. Turkcell in Turkey took customer understanding to a high level through its experience of launching GPRS applications.

Case study: GPRSLand and Shubuo

In 2001 Turkcell launched 'GPRSLand' – a selection of around 20 GPRS applications, packaged and promoted as a mobile theme park with something for everyone. The applications included chat and dating, city guides, a soccer game, a Turkish-to-English dictionary, financial information, news, a biorhythm generator, mobile jukebox, and others. The launch was very successful and Turkcell allocated funding to expand the service.

Analyzing the customer usage of GPRSLand Turkcell uncovered an interesting phenomenon. Although all the GPRSLand services were used by Turkcell's customers, most customers only used one or two of the services each. It became clear that there were 'communities' of customers with specific common interests. For instance, there were customers that only used the music-related services, another group that only used chat and dating, and so on. This behavior is supported by research performed by Ericsson's Consumer and Enterprise Lab, which found, for example, that many people are interested in music, and many are interested in sport, but far fewer are interested in both music and sport.

Turkcell thus launched a new concept, named Shubuo, in which content and applications were grouped into subject categories, Music, Sport, News, Dating, and so on, each of which contained a range of services on the same theme.

5.2.7 Personalization

The Turkcell experience shows how the value of a service is enhanced if it is tailored to customers' needs. We can take this even further by personalizing services for individual customers.

5.2.7.1 Personalization for a individual customer

A simple way to personalize a service is to use the customer's name in some form. This could be included in an SMS message or displayed when he connects to a web or WAP site. He probably knows that it is just pulling a name from a database of thousands, but it shows him that the service has recognized him as an individual and it gives him the confidence to explore further.

A further step would be to display content relevant to the customer's personal circumstances. For example if he has registered his date of birth and place of residence, we might present him with his personal horoscope and the weather conditions in his town today.

The ultimate aim would be for him to have his own personal 'mobile agent', programmed with his interests and needs, that hunts the Internet for relevant content, which it downloads to his phone for instant viewing. However, maybe we should stick to reality for now.

5.2.7.2 Personalization by mood or location

Many instant messaging programs, both Internet-based and mobile, permit the user to broadcast his or her mood using a range of 'emoticons' – or emotion icons. A ☺ indicates happy, a ☺ sad. A *mobile* service has the added benefit of location: mobile positioning technology can locate an individual to within a few meters. So now our customer can see if there are any happy friends nearby who might buy him a coffee. Such a friend-finder combining both location and emotion is supplied by application developer Reach-U and has been a great success for mobile operator Orange in Slovakia.

5.2.7.3 Personalization by interaction

Interactivity introduces a form of personalization as it allows customers to affect the direction or outcome of an application or event. TV SMS voting has proven to be highly popular and there are many examples of interactive SMS and WAP-based quizzes, intelligence tests, personality profilers, etc. You can even date your very own animated virtual girlfriend, who talks to you, has different moods and with whom you must interact by buying her flowers and presents, telling jokes and answering her questions.

5.3 MINIMIZING THE BARRIERS TO USE

We have looked at ways of enhancing our customer's experience, but a mobile experience is not just about enjoying the applications themselves – we must also consider the possible negative experiences our customer may have when using his mobile phone.

Many people are unfamiliar with the possibilities of their phones and use only a few basic functions such as the alarm clock and calendar. They are frequently totally unaware of the existence of online mobile applications and would have no clue how to reach them anyway. We must therefore work hard to make it easy to find applications and to remove the barriers preventing their access and use.

5.3.1 Inside the handset

In one sense the mobile handset is the most important driver of mobile services. It is what attracts customers into phones stores and it is the thing people carry with them practically 24 h a day. It is a fashion item, a status symbol and with features such as color screens, embedded cameras, games, videos and music players, the humble mobile phone has become a veritable multimedia entertainment center.

The handset is both the means of reaching applications and the instrument on which the applications are experienced. The handset is thus the core of the customer experience, and much of the work we can do to reduce barriers centers on this small hand-held device.

5.3.1.1 In-phone applications

All phones today come with a range of embedded content and applications such as ringtones, logos, wallpapers and games. These are quick and easy to find and they always work – it is a really pleasant experience. This is not surprising as the handset manufacturer has selected and tested them specifically for that handset. If all applications were as simple to locate and worked as well you would probably not need to read this book.

The advantage of the in-phone experience is that it eliminates external variables such as data network set-up, transmission delays, communication protocols, portals, etc. So why not bypass these external factors and allow our customer to get content directly to his phone? What if we were to put our content or application on a disk which customers could download to their phones from their PCs using cable, infrared or Bluetooth? There's nothing stopping anyone doing this, and in fact you do not need to be computer scientists to adapt content freely available on the Internet to mobile formats. It is just that most of us do not bother. On a commercial basis we could sell in retail stores CDs with a selection of images, ringtones, videos and games on a particular theme, such as a movie, music artist or TV program. We would have to include all the formats for different handsets, but this is standard practice with mobile portals today.

Music stores are already moving in this direction, with digital music stations where customers can select a range of tracks and either burn a CD or transfer the songs to their MP3 player, so why not to their mobile phones too, complete with a CD cover wallpaper and ringtone versions of each track? Mobile network operators might not like this as it bypasses their main source of income – the network. But then, it is a free market and there is nothing stopping them doing it too.

5.3.2 Escaping the handset – handset configuration

Eventually our customer will decide to leave the comfort of his mobile handset and enter the big wide world, otherwise known as the mobile Internet. Here he has the possibility to download content and applications and explore the wonders of online portals; but wait – it is not that simple.

In the ideal world this section would not be needed in a chapter on customer experience, but as pointed out in Chapter 4, it is very likely that our dear customer's handset is not yet configured to download content. When he walked out of the mobile phone store with his shiny new phone he could immediately make phone calls and send SMS text messages, but if he tries to send an MMS or connect to Internet services the phone will most likely display a message such as 'To access Internet you need to create a profile. Create now?' If he is naive enough to press 'Yes' he will be confronted with requests for an APN, IP address, WAP profile, and more, if he gets that far. This is not a good customer experience.

New phones are frequently not configured at purchase for online services such as GPRS, WAP and MMS. The problem arises because the configuration parameters for data services are programmed into the handset itself, and mobile operators often have no opportunity to load these parameters prior to the phone being sold, either because they are sold through independent stores or because they do not have the facilities, resources and time to program large numbers of handsets from different manufacturers. In contrast, the parameters for voice and SMS are stored on the SIM card, which the mobile operator does have control over. This is why voice and SMS always work even on a new phone.

A few operators, such as Vodafone, have the size and influence to persuade handset manufacturers to preconfigure their phones before they leave the factory, but the majority of operators are unable to commit to sufficient handset purchases to do this. Some operators have their own retail stores with trained staff who can configure handsets and demonstrate services before the customer leaves the shop. For the remainder there is ADC (automatic device configuration), as described earlier. Configuration parameters now filter down directly to the phone the first time it is powered up. Our customer may not even be aware that his phone is being accessed ... and another barrier disappears.

5.3.3   Ordering content

5.3.3.1   Via SMS

There are many ways of getting content to the mobile phone. Perhaps the simplest is to use SMS. Here, an operator or independent content provider provides a range of content items each with a unique keyword. Our customer simply has to send the keyword via SMS to a specified phone number and the content item is either pushed via WAP directly to his phone or he receives an SMS containing a WAP link, which when clicked initiates the download.

If possible the phone number should be a short number and easy to remember. One independent content provider uses the number 35050 and runs TV ads with the tag line 'three-fifty-fifty'. Content can be promoted in a catalogue, or individual items can be promoted in newspapers, leaflets, TV, billboards, SIM vouchers, etc.

Case study: MTN 'please call-me'
One inspired way of driving content downloads is from MTN in South Africa, who have a service called 'Please call me'. South African mobile users can send a USSD string (i.e. they press some buttons) such as *123* followed by the phone number of a person they want to be called by. That person receives an SMS with the message 'Please call...' and the phone number of the sender. It is a way of opening up the mobile market to people with very low disposable income, as it allows people to get others to phone them.

Now here's the clever bit. MTN tags the 'Please call...' message with a content keyword. The receiver of the SMS (who, note, is the one with the money) can now send the keyword to an MTN number and receive a joke or cute message. Are you following this? Let's take an example.

My phone number is 012345678.
My friend's number is 033445566.

With my phone I send *123*033445566.
My friend receives the following SMS:
'Please call <u>012345678</u>. Get a CUTE MESSAGE to forward to your friends! SMS
    CUTE to <u>083123000</u>. Cost R1.50'

This has generated a huge number of content downloads at no additional cost to MTN.

5.3.3.2 'Hybrid' applications

A new approach has been taken both by operators and independent service providers to
use the application capabilities of mobile devices to create a better user experience. For
instance in Chapter 4 we discussed the benefits of the Eniro Yellow Pages Java application.
A trend in the industry is the increased use of downloadable or preinstalled applications
on the mobile phone, which interact with online services 'behind the scenes' and present
what to the user appears to be in-phone information and services. The applications profile
the services to better suit the 'desktop' of a phone, and improve the usability of the
services themselves.

Case study: Mobile Headline Services
One notable effort is from CNN and Reuters together with developer Handmark in their
promotion of news applications. By providing a downloadable Java application that can be
'ordered' through a simple SMS request, users can download a branded news-reader service which gathers the latest headlines and presents them in an easily read short summary
of the latest breaking stories with image content.

Such an application is easily accessible from either the desktop of the screen of from
within the applications folder of the phone and does not require any fiddling with bookmarks or address entries. The experience is also full-screen, so the service provider can
take the opportunity to provide a fully branded experience to the customer.

In Chapter 1 we looked at the gambling area and its move into wireless with an example
from Mansion, who also use a downloadable application to enhance the customer experience.

Case study: Mobile gambling that is fun to use
When Mansion started designing services for mobile, they wanted to reach as many
players as possible but also ensure that the mobile gambling experience was great. For
example, the sports betting application needed to be very quick in its responses, as users
look around the different games to bet on. If it took several seconds to go from one game
to another, the uptake and usage would be lower. Consequently, Mansion needed a rich
and intuitive interface, with instant responses to user navigation and a clear overview
of where they were. An important decision needed to be taken – should they build a
downloadable client or use the WAP browser in the phone? Even though Mansion found
that WAP was good for the purpose, they concluded that the service looked different
on different phones when they used the full potential of WAP. Therefore, a common
denominator of (fewer) features would have to be used to make it work the same way
on all WAP-capable phones. While this WAP service has also been launched, Mansion

Figure 5.1 Mansion sports betting application, four clicks and 2 s to bet on a Major League Baseball game.

decided to also develop a downloadable client to provide a more exciting user experience for high-end phone users. This solution was based on a solution from Zenterio, a company that develops user interfaces based on research into cognitive behavior (how the brain interprets impressions and experiences). The solution needed quite powerful handsets and operating systems, and therefore was implemented in Symbian (C++), Linux and Windows mobile.

The four screenshots in Figure 5.1 show the sports betting application, which is connected to the Mansion server and takes bets in real-time. The customer can without delay (no loading for each step) click back and forth in the content tree, like Baseball → Major League Baseball (MLB) → Boston → New York. An innovative aspect is that the content most relevant to the user is always presented in the center of the screen in order to minimize the cognitive load.

This customized, hybrid approach to improving the user interaction with applications has its own drawbacks:

- the need to download the application in the first place;
- if changes are made to the application, a new version must be downloaded;
- different handsets require different versions of the applications; sometimes even small things, like the support for transparent colors (which in this case makes the application much more user friendly), make it necessary to create specific versions for some handsets;
- so far, the addressable market is limited to high-end phones with open operating systems.

5.3.3.3 Other innovations

Other ways of ordering content include web sites or dial-up using IVR; even the good old SIM Toolkit sits in most people's phones and is still used by some operators, particularly in developing countries where SMS content still dominates. We should also not knock old technology; in fact, ironically, the good old circuit-switched phone call is being use to supply the latest in mobile video streaming.

Operators are experimenting with providing live video streamed to the mobile phone, not as a data call but by making a simple videophone call. Known as Video Short Codes, the advantages include: extreme simplicity – users simply dial a five-digit number to connect to a selection of video streams; guaranteed quality, because a circuit-switched call does not have to share bandwidth like a data call; and predictable cost because calls are charged like a normal phone call, by the minute, which is understood by all customers.

Another clever innovation is the Ki-Bi™ card. This is an electronic credit-card-sized retail device that both simplifies the process of downloading content and acts as its own marketing tool, since the card can be custom branded by the operator, content provider or reseller and sold in normal retail stores.

The card has a number of pressure-sensitive areas which, when pressed, play a series of screeching noises – not dissimilar to the sound of a modem connecting to the Internet. The customer dials an IVR number on his/her mobile phone, holds the card to the phone and presses one of the buttons. Each button on the card is labeled with a content item, such as a ringtone or Java game. The screeching noises are actually instructions to a server on the other end of the line to download the selected item of content to the customer's phone.

5.3.4   Getting connected

OK so our customer's phone is configured, he's ordered a few content items, which were cool, and so he decides he is ready to surf the mobile Internet. Now he encounters the next barrier: how does he connect? Already mentioned has been the 0, 1, 2, 3 concept, with the '1' representing one-button connect, i.e. the phone user has only to press a single button in order to get online. But which button? Back in 1999 NTT DoCoMo made it simple by having a button on i-mode phones labeled 'i-mode'. Brilliant! For the rest of us, things are slowly getting better: my current phone for instance has a button labeled 'online'.

We are also seeing a continual push by operators to exert control over the mobile handset, including the color and branding as well as the handset user interface. Device vendors now allow limited customization of the menu layout to include shortcuts to the operator's portal, and other features.

As a support to those operators that do not have such strong influence with vendors, as well as for large enterprises wishing to deploy company-wide solutions, this concept has been taken one step further. Several providers allow for a more extensive software customization of the look and feel of key functions of the mobile handset. Through the use of embedded applications, it is now possible to rebrand the desktop of a phone as well as to more actively promote various portal features such as content promotions and previews into this branded experience.

Case study: O2 Active – 'one click' service access
O2 set out to reinforce their O2 Active brand and provide an updateable 'shop window' for content with an improved customer experience. By teaming up with mobile software provider SurfKitchen, O2 were able to launch a comprehensive application-based desktop

Figure 5.2 Custom menu from O2.

solution that was rolled out on a number of smart phone devices – this was called O2 Active Menu (Figures 5.2).

From the user's perspective, the desktop of the phone provides 'single-click' access to key O2 Active content and services, e.g. sports, downloads and more. Within the portal-related menus, popular wallpaper and ringtone content is available on the device for instant preview and these lists are automatically updated over the air as new popular items are added into the portal. This desktop does not preclude normal use of the phone for making calls and messaging, but it provides a full-screen presentation of the O2 Active brand and a highlight of key services.

SurfKitchen's technology addresses the need for mobile operators to create and deliver a diverse range of content-related services through a simple and consistent user interface. It also provides the flexibility to change menus, links and content over the air, allowing the creation of special campaigns and the promotion of important services. This simplifies the user experience, enables easy discovery, encourages adoption and promotes brand identity.

By providing the ability for our customer to browse services off-line, it is clear that he will be more willing to explore new services and try out new content without the fear of having portal access fees sprung on him before a purchase has been made – the equivalent of charging him for just walking into a high street store or expecting him to purchase clothes without trying them on.

5.3.5 Finally online – the WAP portal

Eureka! Finally we have guided our customer through all the pitfalls and he arrives at our WAP mobile portal, which brims with exciting content and applications. But don't relax yet. Even though we have the customer in our domain, we have to help him find something that will persuade him to make a purchase.

5.3.5.1   Fewer clicks, less content, more downloads

It is very hard to make things easy and when you have a small screen and a WAP round-trip delay of a couple of seconds, so do not be surprised if our customer's initial enthusiasm quickly fizzles out. We still have work to do to make sure his experience is such that he is likely to come back for more.

There are countless books covering good practice for WAP page design, but in a nutshell it should be intuitive to use and designed such that the customer can find interesting content as quickly and simply as possible. This generally means having a broad, shallow portal, i.e. one where you can reach content in as few clicks as possible. It is said that for every click, you lose 50 % of your customers. WAP designers have been seen to dance down the corridors when they have managed to eliminate a single click from an application.

Downloads can be increased by having less content. This might sound like a paradox – surely more content means more downloads – but there is plenty of evidence that customers are inherently lazy when it comes to surfing web or WAP pages. For instance, content listed in alphabetical order inevitably means that items whose names begin with A or B will be the most downloaded. It is easy to counter this by simply rotating the order in which content items are displayed, on a daily basis. This guarantees that each day our customer will see an apparently fresh piece of content at the top of the menu, and will more than likely download it.

It is better to have a few items that change regularly than 1000 items that remain static, which the customer has to click and scroll through for hours on end. Also, include a list of top-10 downloads – they do not have to be the actual top 10, but they soon will be because that's what most people will download!

5.3.5.2   Dynamic menus

In Section 5.3.3.2 we looked at handset-based applications that installed a user-friendly menu in the customer's handset in order to shield him from online interactions. There are also solutions that work in the opposite way. A solution from Uniper places a server on the network that monitors each user's behavior and adapts the WAP menu based on past activity. So if our customer consistently selects the Sport item of a menu, Sport will be moved up the menu, so he has less far to scroll the next time he connects.

5.4   SUMMARY

This and previous chapters have illustrated that there is much more to mobile applications than the applications themselves, and that there is probably more effort required to manage a customer's experience of an application than it takes to develop the application in the first place. The customer experience encompasses everything from first hearing about an application to actually seeing it on the mobile phone. Unfortunately the distance between these two events is still great, but it is shrinking. The move appears to be towards putting more functionality in the handset and relegating the network to background tasks, much as computing power transferred from the central mainframe to the desktop back in the 1980s. Let us therefore look more closely at the handset and the power we all have literally in our hands.

6

Mobile Devices – Leading the Way

6.1 THE IMPORTANCE OF MOBILE DEVICES

The mobile device has a pivotal role in determining the success of new mobile applications by providing an intuitive user interface and packaging existing and emerging technology in an attractive format. The device shapes the perception of the user and defines whether users will appreciate new services and start using them.

We call it a mobile 'device' because we should not limit ourselves to thinking of the device as just a 'mobile phone'. Today's mobile device goes far beyond delivering voice services. It is evolving towards a device that can handle virtually all communication needs, using a wide range of media formats. Also, the device itself is able to host local applications, making it into a PDA replacement and a games console as well. Yet the device is just the last link in a whole service delivery chain, a delivery system with capabilities developing at different paces and end-to-end capabilities that define the user experience.

The only constant with the device market is change. Different roles are appearing: platform providers, application providers, device providers/integrators, etc. There are some parallels to what happened when the PC industry was horizontalized 20 years ago, but there are also fundamental differences. The problem symptoms of the early personal and home computer industry are similar: scattered operating systems, interoperability issues and complex diverse user interfaces. However its important to bear in mind that there are some fundamental differences, both when analyzing the two industries and when forecasting their development to make decisions on how to deploy mobile applications.

One difference and paradox in the mobile device industry is that the more sophisticated and 'high-end' the device is, the more it is built on standard components. One could expect low-end devices to be built on standard components to reach economies of scale, but this is generally not the case. Instead typical high-end devices use a multivendor operating systems such as Symbian and Microsoft Pocket and include standard software packages like Realaudio and MS Mediaplayer to provide powerful applications and media capabilities.

Mobile Media and Applications – From Concept to Cash: Successful Service Creation and Launch   Christoffer Andersson,
Daniel Freeman, Ian James, Andy Johnston, Staffan Ljung
© 2006 John Wiley & Sons, Ltd

Low-end devices are built on proprietary operative systems and deploy vendor-specific graphical user interfaces. The key driver behind this difference between high-end and low-end is, not surprisingly, price. A standard operating system has a cost of US$8–15 while a proprietary solution cost less than US$1.[1] We should remember that the lion's share (some 70 percent in a typical market) is low-end phones – there is a constant battle to reduce the bill for materials between the device manufacturers. In this war US$10 is a lot.

Another important technology difference between PCs and mobile devices is the constant need for power and space optimization in the latter. This has resulted in a tight integration and optimization of the device components rather than the use of standard components with open interfaces like in the pc industry. The flip side of this is compatibility issues between devices and a low level of commonality between different mobile devices. Another important difference to the PC industry is economies of scale: in 2004 alone, more than 700 million mobile devices were shipped – more than the total installed base of PCs.

Wide area network access – the very core functionality of the mobile device – is of course another differentiator. Evolving from analog circuit switched voice to multiple access technologies that support higher data speeds enables more functionality but also poses new problems for mobile applications and challenges to understand what is really applicable. Entering into the next phase of radio interface technology, WCDMA radio illustrates this well. Of course, this has profound effects on the performance characteristics of the bearers available, which means some applications can move from the realm of experiments to mass-market-ready services.

The experiences from the introduction of GPRS and WAP services have taught us that easy service provisioning and configuration is a prerequisite for service uptake. For many applications providers this is still a key issue – just to count on the mobile operator to provide a plain GPRS/WCDMA connection as a mobile Internet Service Provider (ISP). The explosive growth of services and content that the new technologies make possible must be available with essentially zero configuration work by the user.

Price, ease of use and coverage are other basic requirements for an end user, since the loss of functionality of a mobile phone is devastating to the consumer. The basic communication functions of the phone (voice, messaging) are of such importance than any risk of tampering with these will deter the user from using more advanced functions.

Furthermore, it is of utmost importance to avoid inheriting the vulnerabilities of PCs and the Internet. In particular, protection against viruses, trojans, and general forms of malicious software is a key design criterion for mobile device manufacturers. Secure execution environments, signed code download, etc., are functions that several mobile device forums have worked intensively to resolve and have come a long way in resolving.

To analyze the limitations and emerging mobile application opportunities in more detail, we will go under the hood of the mobile device and see the impact of its building blocks in Section 6.2. We take this one step further in Section 6.3 where we see how the functionality evolves and diffuses from high-end niche phones to become commodity features we can count on to build mass market applications. After this we go beyond the device itself and look into some enablers that facilitate the creation of cross media experiences. In

---

[1] Ericsson Mobile Platforms.

Section 6.5 we relate the possible changes and improvements in the building blocks to the emergence of the main fields of applications we have seen. Fields of applications are driven by the advancement of the mobile device technology we reviewed in the earlier sections but require much more to become mass-market products. Following this, Section 6.6 concludes with a discussion of mobile device adaptations of the different levels.

6.2 MOBILE DEVICE ARCHITECTURE – THE BUILDING BLOCKS

6.2.1 The components under the hood

We base our break-down of the key components of the mobile device on a simplified open system interconnection (OSI) model. This is a layered model going from low-level network communication to the mobile applications we use. It is a widely used model in different fields of telecommunications that helps us to structure the explanation both to understand what the components are and what they do. The ultimate goal to understand how all the 'techie stuff' impacts our ability to create better applications and media experiences (Figure 6.1). On the highest level we have the operator or customer applications. The capability of those applications is ultimately defined by the functions that the lower levels provide to the application developer. So let us go down to the lower levels!

In this world of acronyms and 'details' that indeed can change the user experience from thrilling and boring, the key question we ask ourselves is: how will all this impact the ability to create more attractive applications and use the mobile as media channel?

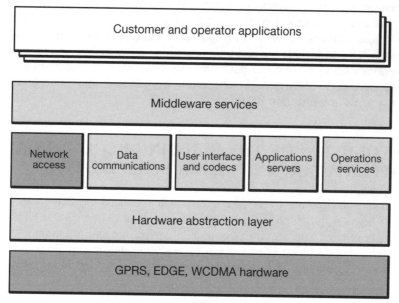

Figure 6.1 An OSI view of modern mobile device. Source: Ericsson Mobile Platform.

The layer below the applications is the middleware level. It is an abstraction of all the functionality that can be accessed by the mobile media applications. Here is where we find the Java execution environment (or BREW) and a varying set of APIs, for example the APIs defined by the Java standard.

If we now go down one further level we find five building blocks of somewhat related functionality: network access, data communication, user interface and codecs, application servers and operation services. The network access layer provides network access to GSM, GPRS, EDGE (enhanced data GSM environment); WCDMA (CDMA DV/DO for devices operation in those standards). The complexity and the challenge of obtaining the required stability and functionality have increased over time. This is one strong driver behind the emergence of mobile platform suppliers whose platforms are used by many mobile device manufacturers. To put a figure on this, in 2001 a plain GPRS phone had 300–400 requirements, while an advanced 3G phone in 2005 has 6000 system level requirements.[2] The required economy of scale has led to the ongoing restructuring of the mobile device business with fewer platform vendors. To analyze future functionality (the next 1–3 years), it pays to look at the functionality of these platforms since there is a lead time of some 12–24 months from the time of the platform specification to the time when the devices built on these platforms are available in the market.

Coming back to the user interface and codecs, this is functionality related to the input and output devices of the mobile. This includes handling of cameras, screen displays, keyboard and touch screens. We recall many protocols that we take for granted in the Internet world, protocols that enable richer audio, video and image experience in compressed formats (for example AAC, MP3, WB-AMR, H.263 and MPEG-4). Some of them require support in hardware to run really smoothly, but since there is a cost associated with this, hardware acceleration is only deployed in high-end phones. A *hardware abstraction level* is introduced to make the lower level hardware transparent to the software layer.

Data communication services include personal area network connectivity either wirelessly by infrared or Bluetooth or fixed by a universal serial bus (USB) or serial cable (RS232). It also includes the support for Ipv4/Ipv6 and enablers for synchronization.

Operations services include the real-time operation system and provide important low-level functions such as battery and power management, memory handling, file management, platform security, SIM access and control. We will look into a couple of different operation systems in the section below.

6.3 EVOLUTION OF THE BUILDING BLOCKS IN AN APPLICATION AND MEDIA PERSPECTIVE

Today's mobile devices are evolving quickly, with enhanced capabilities to play back music, take high-resolution photos and watch short videos and broadcasted TV. A sound understanding of what the capabilities of mobile devices are and of the underlying technology driving the evolution is paramount to exploring the opportunities and avoiding the pitfalls in this business. This rapid evolution of multimedia capabilities is only partly

---

[2] Ericsson Mobile Platforms.

related to the introduction of 3G access bearers. The real driver stems from the widespread audio and imaging applications and advancement in technology that allow for low-cost miniaturization. Let us now explore some key components that impact the success of mobile applications.

6.3.1   Operating systems

Low-end devices use proprietary operative systems and deploy vendor-specific graphical user interfaces. The trend is to move away from vendor-specific operating systems (OS), but since the cost for a high-end OS is still US$8–15 compared with less than US$1 for a proprietary low-end OS, this is only happening in mid-tier and high-end segments.

Smart phones have an operating system that is open to application developers. They are feature-rich and typically have a large high-resolution display and lots of storage capacity. However smart phones have a high bill of material (BOM) cost, which leads to relatively low volumes and high prices. We expect that more and more of the devices with proprietary OS will be replaced with devices using an open OS. This will make the mobile device market less scattered over time. This trend to use standard OS is accelerated by the rapid fall in price of memory and higher processor capacity. To what extent smart phones will be adopted and not remain a niche is disputed; an estimate by the consulting company Shostech is that 20–40 % of phones will be smart phones by 2007.[3]

Gradually the functionality of open OS will be diffused to the mid-tier and low-end phones as well, which is why we need to study OS development to understand where the mass-market device functionality is heading with a 2–3 year delay from when high-end phones are released to the market.

The dominant OS for smart phones is Symbian, with an estimated market share of 61 % of the smart mobile device market, followed by Microsoft smart phone OS (18 %), PalmSource (10 %), RIM (7 %) and others (3 %).[4]

So what is the OS responsible for? The Symbian and Microsoft Stack can be abstracted similarly in that they have three layers: the core operating system, the user interface and the execution environment (see the left-most side of Figure 6.2).

The operating system provides the core functionality of the phone (see the right-most side of Figure 6.2). The circuit switched connectivity provides the 'telephony' capabilities of different standards (GSM/EDGE/WCDMA/CDMA2000). The multimedia component includes frameworks for plug-in cameras, speech recognition, video playback and recording, etc. Networking includes the HTTP transport framework. Bluetooth and irDA provide local connectivity protocols. Web provides rich text rendering with auto-sizing text editor recognition of URL and e-mail addresses, among other things. The messaging component for Symbian has functionality for exchange of business cards and meeting information (Vcard and Vcalender). It also has ringtone functionality and settings for Internet access. Graphics has the functionality for drawings (rectangle fill, polygon drawing) and direct screen access. It accesses acceleration hardware through the hardware abstraction layer.

---

[3] The Shostech Group 2004, 'Device convergence, three worlds coming together'.
[4] Canalys 2005, 'Market shares by operating systems Q1 2005 for the worldwide total smart mobile device market'.

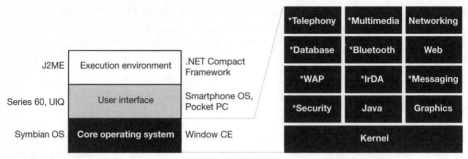

Figure 6.2 Overview of the OS and its functional components. Source: Symbian, Ovum and Ericsson.

Java provides the Java execution environment that will be discussed more in detail in the following sections. Database includes synchronization functionality (SyncML client for Symbian) and database adaptors. The kernel has basic file server functionality, user library and device drivers.

On top of the core operating system there is a vendor-specific implementation of the user interface such as Nokias Series 60 or the UIQ extension for Symbian for pen-based phones such as SonyEricsson P900. The Series 60 Platform, built on the Symbian OS, is currently the leading smart phone platform in terms of sales. It is licensed by some of the foremost mobile phone manufacturers, including LG Electronics, Lenovo, Nokia, Panasonic, Samsung, Sendo and Siemens. UIQ phones use large, touch-sensitive screens (208–240 × 320 pixels, 12- or 16-bit color). The latest release brings advanced user interface personalization with themes.

6.3.2   Java – software enablers for mobile devices

Why is Java important to the application developer and how does it enable a richer mobile media experience? Java and its Java 2 Micro Edition (J2ME) is the most widespread application environment. Java enables applications to run on your mobile. It has a large community of developers and the code is easy to port and reuse on any device, even though in reality most advanced applications still need modifications to run well on different devices. An illustration of some typical Java applications is shown in Figure 6.3.

The attractiveness of Java is that it has a widespread developer community that allows for ubiquity. The installed base is growing from almost 900 million devices in 2005 to an estimated 1.5 billion in 2008. BREW is an effort by Qualcomm to provide in particular CDMA (code division multiple access) operators with an end-to-end environment for mobile applications. Hence BREW is less widely used in Europe than in the USA and Asia, where we find more operators using CDMA. BREW also supports Java and, if we add BREW to the picture, we will have almost 2 billion enabled devices by 2008 – this is a global mass market![5]

---

[5] Ovum Global device forecast 2005.

Wireless Gaming
The richer and colorful graphics and media capabilities that J2ME supports can provide great opportunity for building feature-rich mobile games

Travel maps
Java provides a means to create visually appealing map-based solutions that leverage the in-built functionality of the phone

Persionalized service
Java and the related websites enable the users to obtain personalized management Java based applications such as health calculator, expense tracker etc.

Location-based service
J2ME with the location package facilitate the availability of more location-based applications with more attractive visual effect and more user-friendly interface

Mobile music
Java with media support and graphic support enables the development of more powerful mobile music applications which can offer music/ringtone download, wallpaper and music video services

Figure 6.3 Sample Java applications. Source: Ericsson 2005.

A large target market is of course a key criterion for successful media applications; however, as we will see, there are trade-offs between ubiquity and functionality. So the Java slogan 'write once and run everywhere' (PC, PDA, mobile device) is not entirely true since the APIs supported define the key functionality and vary between the different editions of Java (J2EE, J2SE and J2ME). For devices with limited memory (less than 512 kbytes), such as mobile devices, there is a special configuration of J2ME called Connected Limited Device Configuration (CLDC). The specific mobile traits are defined by the Mobile Information Device Profile (MIDP), which has widespread industry support. These are defined by Java specifications (JSRs), which we will come back to and highlight later in this section.

MIDP should be regarded as a common core of agreed APIs that any device supporting MIDP has. So what is the evolution of MIDP in terms of ubiquity and functionality? MIDP has evolved from MIDP 1.0 to MIDP 2.0. It is backward compatible so a MIDP 2.0 phone will run Java programs developed for MIDP 1.0. CLDC has evolved from 1.0 to 1.1 and most MIDP 2.0 implementations will use the latter version. Here is a summary of what MIDP 2.0 brings:

- *Uploading enhancements* – in MIDP 2.0 the over-the-air loading is standardized. MIDP 2.0 also supports code authentication and signing using digital formats.
- *Access control and security mechanism* – controls access to sensitive data such as the address book and supports an end-to-end encrypted pipe.
- *Enhanced user interface* – pop-up choice group showing the current selection, interactive alert screens with gage, form formatting by providing layout directives, customizable components of a form, etc.
- *Improved Audio and image support* – includes audio building blocks to add tones, tone sequences and playback of WAV files in the application. Supports RGB images, which enables manipulation of picture integer, arrays and pixel-by-pixel transformations.

The uploading mechanism is especially important since there have been many issues with problems of loading the application and acknowledgement that the application has actually been installed successfully. This will further make Java a basic enabler for games and interactive applications, which will drive adoption of games in the enabled base. In an estimate from Analysys,[6] adoption of Java games had already surpassed WAP/SMS games in 2003/2004. While the former show a rapid adoption and will reach some 25 % of the Western Europe population by 2009, the latter will only gain slowly and will stay at an adoption rate of around 5 % in the same year.

Even though Java is driven by standards, the reality is that most device manufacturers and many game platform vendors have developed their own additions to Java to optimize the use of particular devices. For example, RIM's Blackberry includes an extensive set of nonstandard Java APIs that allow for e-mail connectivity. The implementation of the different JSRs varies across devices. We have already touched the core specification of MIDP and CLCD. To obtain a more complete picture we have to look in the 'umbrella' JSR 185, known as specification of the Java Technology for the Wireless Industry (JTWI), which gives roadmaps of supported functionality. JSR 185 is certainly a step in the right direction. It will make device functionality more coherent over time, but still the degree and version of the deployed JSRs varies. Because of this diversity of support of JSRs (and even versions of JSRs, e.g. JSR 135 1.0 and 1.1), it really does make Java *not* write once, run everywhere.

Table 6.1 gives an overview of the key JSRs. For example JSR 120 allows for SMS/MMS and other messaging features – a key factor in making simple interaction possible. All the functionality of JSR 135 is important (note support for 1.1) in allowing access to multimedia playback – without this you cannot stream content or play back stored material within the application. This is key since developers can rely on the multimedia features of the device without having to create their own!

Table 6.1 Overview of different Java specifications that standardize key features and APIs. Source: Sun Microsystems and Visiongain 2004

JTWI	JSR185	Roadmap
CLCD 1.0	JSR30	Mandatory in JSR185
CLCD 1.1	JSR139	Conditional in JSR185
MIDP 1.0	JSR37	Basic specifications for early Java version phones
MIDP 2.0	JSR118	Mandatory in JSR185
Mobile media API (MMAPI)	JSR135	Conditional in JSR185
Wireless messaging API 1.0	JSR120	Mandatory in JSR185
Wireless messaging API 2.0	JSR205	Enhanced messaging API
Mobile 3D graphics API	JSR184	Important for games, gives 'console' features
Location API	JSR179	Defines how location services are accessed
Security and trust services API (SATSA)	JSR177	Security and trust API
J2ME web services	JSR172	Defines web service access

---

[6] Analysys 2004, 'Making a success of the mobile content value chain'.

The implication of this is that a game developer that would like to make their games available for some common handsets would have one version for the Nokia Series 60 MIDP1 version (N-Gage, 3600, 3650, etc.), a generic MIDP2 version (Nokia 6600, QTek S100, Sony Ericsson P910i, Sony Ericsson K700i, etc.) and a Nokia Series 40 version (3100, 3300, 6800, 7250, etc.). For the Nokia Series 40 version, some of the smaller details would have to be removed from the game, in order to fit within the limited space available on such devices. There could also be an additional MIDP2 version, but designed to fit smaller screen sizes (Sagem myX5-2, etc.). How is this done by a large games publisher like Mforma?

Case study: how Mforma builds games based on different mobile device capabilities[7]

MFORMA is a leading global publisher and distributor of mobile entertainment. It has a large catalogue of mobile games and connected applications for JAVA, Brew, SMS and WAP platforms as well as music, lifestyle, and sports subscription products. MFORMA has established key partnerships with leading entertainment companies such as Marvel Enterprises, from whom it has licensed the rights to mobilize Marvel's entire catalogue of world-renowned comic characters such as Spider-Man, The Fantastic Four, The Incredible Hulk and X-Men.

One of the challenges to mobile entertainment publishers today is caused by the great number and diversity of devices in the marketplace. MFORMA currently delivers content for more than 300 different devices, each with different capabilities (screen resolutions, memory, 3D graphic support etc) and versions of software. In line with leading industry practice, MFORMA typically designs a game based on the capabilities of the top-end mobile devices. They then create as many as seven different reference builds in order to ensure maximum quality and playability for the typical devices. To do this they go beyond the standards and use device-specific functionality. After having produced the top-end versions, they engage dedicated porting resources to build additional versions. MFORMA has porting teams in San Francisco, USA, as well as in the UK and China. These porting teams specialize in adapting graphics, user interfaces, etc. to all variants of devices and guarantee an optimal output across a very wide range of devices.

6.3.3   Processor capacity and enabling chipsets[8]

Looking under the hood of the mobile we find its brain and heart. The brain is the CPU that handles general-purpose software, including the GUI, and runs the TCP (transmission control protocol)/IP communication stack. The digital signaling processor(s), DSP(s), handle the radio protocols. In this analogy, the heart, which is what the mobile uses to sense and reproduce feelings (well, we mean ringtones and small blurry pictures), would be the sound chips, camera sensors and graphics accelerators.

The most important enabling chipsets we see in an applications view are:

- the CPU (the brain of the mobile) itself and the DSPs (the lower level communication handler);

---

[7] Source: Mforma.

[8] The main sources in this paragraph are Carter L. Horney, 'Global cellular hand set and chipmarkets 2005 and Ericsson Mobile Platforms.

- ringtone and sound chips;
- camera sensor and camera chip;
- mobile graphics coprocessors;
- applications processors.

Virtually all mobiles have a DSP and an CPU. Apart from handling the base band protocols they are also increasingly capable of generating ringtones without a separate chip. MP3 is often played on the CPU, but support routines in either the CPU itself or one of the DSPs is used to speed this up. Some phones and PDA have full MP3 decoding/encoding in hardware as it is 'cheaper' battery-wise to run MP3 in hardware than a mixture of hardware/software. To put a number on processor evolution, SonyEricsson's Z1010, which was commercially available in 2003/2004, was built on a platform that had a processor running up to 104 MHz, while the next generation UMTS/GPRS/GSM hardware platforms that are available in 2005/2006 have doubled this. When comparing the first- and second-generation UMTS/GSM platforms, taking into account CPU speed, cache and memory performance, the CPU performance is around 1.5–1.7 times faster.

What is happening with the chips that makes possible the biggest mobile applications by revenue – ringtones? Qualcomm has chipsets for Compact Multimedia extensions (CMX), which is capable of MIDI, SMAF and other sound formats. It also has the capability to play music that is synchronized with picture (PNG/JPEG graphics) and text. It can handle 44 kHz sampling rates in stereo, which is certainly a different experience from the once so simple monophonic ringtones! To give an idea of the volume, 55 million devices were shipped in 2004 for the CDMA market with this capability. The largest audio DSP chips manufacturer in the world, Yamaha, has been a driver behind the 'Synthetic Music and Mobile Applications Format' (SMAF). This standard is used for creating ringtones for mobiles but it also supports display of text and graphics.

Going from sound to images, let us look under the hood of the mobile at the components that make picture capture possible. This is the real hit of the last 2 years and one of the most successful extensions of consumer electronics ever. The key component for the camera capability is the sensor. There are two types of sensors, CCD (charged-coupled device) and CMOS (complementary oxide semiconductor). The latter dominate the mobile camera market since they are cheaper and have lower power consumption. Over time the CMOS technology has been improved and is close to that of CCDs. The latest CMOS can handle up to 3 Mpixels. According to Sharp, together with Sanyo Electric the dominating camera module suppliers for mobiles, high-end phones are likely to use two sensors: a high-resolution CCD for photos and CMOS sensor for auxiliary display functions. These camera modules also include color processor chips with auto-exposure and auto white-balance control to enable photographing in different environments: day and night with and without flash.

The increasing demand for the mobile to handle graphics and multimedia drives vendors to include graphics coprocessors. Typically they will be embedded with the DSP/ARM boards. Newer chips (such as the mobile multimedia processor MB86v01 from Fujitsu) have functionalities like 2D/3D graphics accelerators, handling of two cameras, display of multiple images on separate screens (for flip phones), mixing of multiple visual images using transparency features, compressing and decompressing video (MPEG-4) and displaying 15 frames/s, compressing images, and generating 27 Mpixel/s in graphics display. This particular chip was released 2004, which means that it is in high-end phones in 2005 and similar capabilities will be seen in mid-tier phones by 2007.

In conclusion we see the multimedia capabilities develop quickly and diffuse from high-end to low-end in a cycle of around 2–3 years. It is important to recall that, for many applications such as Java games, the typical device is mid-tier to low-end and, as an application develops, one needs to hold the horses for a while before all 'exciting' features are really part of the target market.

6.3.4   What the base interface and network access give

Network access technology defines the capability limitations of providing connectivity to the wireless device – to interact with your mobile wherever you are. This is not a book on network standards and technologies, but let us review the network access types from a device perspective and purely in terms of what they provide to the end user. The air interfaces, i.e. how the mobile device connects and exchanges information with the network, is defined by different standards. Different standards have different characteristics. The key indicators from a user's point of view of these standards are:

- *Global reach* in terms of number sold mobile devices and geographical coverage. This indicates the ubiquity and the economies of scale. A large global reach leads to lower mobile device prices, which is a prerequisite for mass-market penetration.[9]
- *Speed under realistic conditions*, i.e. what you as a user of the air interface actually can expect to get, not what the standard states you could get. Chapter 4 also explored ways to design applications taking these realities into consideration.
- *Latency*, which is the responsiveness, i.e. how long does it take for a request to be handled by the system in terms of delay, the message round trip time. This is important for the applications developer of online games, TV media interactivity and other 'chatty' applications to be aware of. In Chapter 4 we reviewed the design implication of this.
- *Quality* can be measured many ways, but to make it simple we use a relative measure to something you can relate to, namely what we have today.

To simplify, there are two main important tracks of standards development, CDMA and GSM/WCDMA. There are others, such as TDMA and iDEN, but they will evolve into one of these tracks. Figure 6.4 highlights the development of the key aspects.

In terms of volumes, GSM/GPRS/EDGE dominates, with more than 60 % of the market share of mobile device shipments in 2004. It is the dominant standard in Europe and also a data migration path for many North and South American operators that have previously used TDMA. CDMA is second largest with 20 % and is the dominant standard in the USA. The Japanese 2G technology, PDC/PHC, still has 9 %, but will migrate to WCDMA over time. The WCDMA standard for Japan will merge with the WCDMA standard outside

---

[9] One can argue that global reach is not a end-user indicator, but we will use global reach as an indicator of high volumes, high-level competition and good coverage, which we assume translate into good value for money for the end-user.

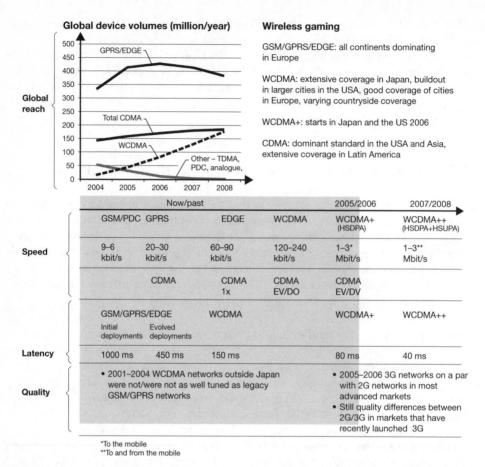

Figure 6.4 Development and comparison of key aspects of different network standards. Source: Ericsson and Ovum 2005.

of Japan by the end of 2005. This will help WCDMA reach 200 million devices by end of 2008.

One important difference between CDMA and GSM is that CDMA has less diversity of suppliers of chipsets to the device devices. The dominating technology provider is Qualcomm[10] for CDMA, while for GSM/GPRS/EDGE/WCDMA there are many suppliers. This has led to a quicker standardization of the interfaces and an initially quicker migration to its 3G standard CDMA EV-DO. The GSM path has been slower but has on the other hand more industry support, which will render more economies of scale long-term.

An important industry development in the device business is the modularization and specialization of work. There are a few mobile device platform suppliers that have developed

---

[10] Eighty-five percent of the CDMA chip market according to Carter L. Horney, 'Global cellular hand set and chipmarkets 2005'.

the base interface that provides network access used by multiple device manufacture's that integrate and manufacture the device sold to customers

For WCDMA there are different radio access bearers (RAB) that define the uplink and downlink speed. The mobile device (and network) support for multiple RABs defines the possibilities for multimedia multitasking scenarios. See Chapter 4 for a discussion on how this impacts the design of mobile applications.

For GSM there are packet switch and circuit switched connections. An important enabler is dual access transfer mode (DTM), which means that the phone is capable of handling both a voice call (circuit switch) and a packet switch connection.

6.3.5   Impact of Bluetooth, WLAN and other local area access technologies on mobile applications and media

Short range technologies provide hot spot coverage (WLAN) or personal area networks (PAN) where files can be exchanged with others. The promise to the user is cheaper transfer of files and high-quality low-cost voice-over IP. But will this happen? And what will it mean to the applications developer?

Bluetooth has taken longer to be adopted than the technology evangelists thought a couple of years ago. The advantage is that Bluetooth chipsets are cheap (a dollar or two) and that they have a protocol stack that makes information exchange easy – it has ready-made profiles for earphones, printers, and sending and receiving of contact details. Contrary to the early standards of WLAN, it has good protocols for high-quality voice transfer. Despite the rich protocol stack of Bluetooth, it has had difficulties in obtaining user adoption beyond headsets, even though it made it under the hood of the mobile device – people find it easier to plug in a USB cable than activate Bluetooth and connect to another device. Strategy Analytics[11] have quantified this and estimate that USB will be available in 40 % of all devices shipped globally by 2005. This will go up to around 75 % by 2010. The same number for Bluetooth is 10–15 % lower but follows the adoption rate of USB.

So what applications will be used based on Bluetooth? British Telecom is using Bluetooth to connect its 'Blue phone'. It is a convergent offering where the user connects with a Bluetooth access point at home, which then uses the normal broadband TCP/IP to provide voice over IP. Outside of the PAN the device accesses the normal GSM network. In the convergent devices Motorola supplies, Bluetooth was chosen over WLAN supplies due to lower power consumption, but also for better voice quality.

Nokia is making an effort to drive the Bluetooth adoption beyond headsets with its the 'Nokia Sensor'. It is phone software application that offers a new way for people to create information and share it with other phone users nearby. Users can create personal pages on their phone, including text and graphics. Nokia Sensor users can also check out the pages of other Sensor users in their vicinity, exchange messages and share files with them.

What about WLAN? We have read press releases for some years of the threat of WLAN, but what is the reality? Convergent GSM/GPRS/UMTS/WLAN phones have finally made it to the market in more than laboratory test volumes. In July 2004 NTT Docomo launched

---

[11] Strategy Analytics 2005, 'Taking camera phones into digital camera territory: Megapixels, WLAN and Printers'.

its N900iL WLAN/3G phone weighting only 120 g and allowing more than 200 h WLAN connectivity. Nokia and Motorola, among others, also launched convergent WLAN phones in 2004.

The gaming industry especially and its new hand-held devices like PSP have adopted WLAN and this will drive down chipset prices and increase adoption. The promise is high speed at low cost; the question marks are ease of use and ubiquity. However, most likely convergent 2G/3G/WLAN will remain a niche product for the next couple of years.

Any short-range technologies with a substantial footprint in the market have a roadmap for higher speeds and better quality. WLAN and Bluetooth have also increased their range over time. Bluetooth may use the so-called ultrawide band (UWB) technology to evolve beyond Bluetooth 2.0's capability of 2.3 Mbit/s. UWB is primarily designed for TV connectivity to set-up boxes and connectivity between DVD and video servers with up to 200 Mbit/s transfer rates. WLAN 801.11b, with a maximum application throughput of 5.5 Mbit/s, has evolved with 801.11a,g that promise up to 25 Mbit/s. However, perhaps the most important development from a mobile device perspective is the enhancement of voice handling in 801.11e. It extends the protocol with voice handling capabilities. However it remains to be seen what the industry support and commercial adoption will be. There are other promising technologies such as near-field communication (NFC) that enable secure exchange when the mobiles 'touch' each other, which is for example used in NTT Docomo's payment service 'Felica'. It has large device support from giants like Nokia, Samsung and Sony. Zigbee have focused on niche applications such as control of lightening, heating and cooling systems. Figure 6.5 gives an overview of this. The development continues but let us first wait to discuss those until the existing technology makes it beyond high-end and niche phones.

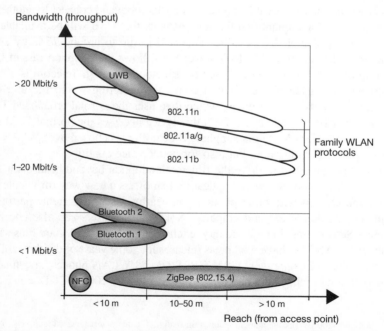

Figure 6.5 Overview of short-range wireless technologies. Source: Analysys 2005.

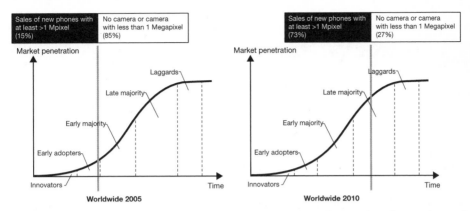

Figure 6.6 Multi-megapixel cameras move into the mass market. Source: based on estimates by Strategy Analytics 2005.

6.3.6 Camera and image capture

We have seen the evolution of the enabling camera chips, but what has happened on the market? And what can we learn from the camera phone introduction when it comes to diffusion from low-end to high-end? The rapid introduction and adoption of camera mobiles surprised even the device manufactures. In 2004, 257 million camera phones were shipped, which was 38 % of the total number of devices in that year. The year before, the ratio was a mere 4 %, which shows how quickly the market is developing. In just a couple of years built-in cameras have become a must-have feature of mobile phones. In 2005 slightly more than half of phones sold will have a camera; out of these one-third will have multi-mega-pixel capabilities. By 2010 almost three-quarters of the phones sold will have this capability (Figure 6.6). If we map this to the market adoption curve, we are on the verge of entering the early majority.

Apart from the multi-megapixel functionality, camera phones also contain several important features to handle the pictures. This includes brightness, contrast, color, zooming, cropping, rotation and overlay. These functions are enablers for creating a richer and more fun experience for the user, e.g. editing an MMS before sending it. The real-time processing of megapixel resolution images is demanding and often requires hardware acceleration to handle compression and decompression. Multi-Mega-pixels sensors, flash, autofocus and optical zooms are standard in digital still cameras and will over time be included in camera phones as well.

This development is important for the applications developer since the rapid mass market adoption of camera phones will accelerate the adoption of terminals with increased storage space and graphics accelerators that can be used for games and other interactive applications.

6.3.7 Battery and power consumptions

One of the limiting factors for mobile devices is the battery life of the devices. The importance was proved in several customer surveys that showed that one of the major

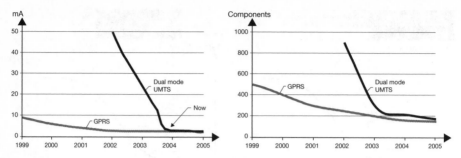

Figure 6.7 Power consumption of GSM and UMTS dual phone over time. Source: Ericsson.

customer complaints when both GSM and 3G were launched was the limited battery time. For 3G, this was due to high power consumption in the initial implementation of dual UMTS/GSM devices since the chipsets were not well integrated and optimized. The consumption is related to the number of components in the device. When the chipsets become more tightly integrated this translates into fewer components and less power consumption (Figure 6.7). The mobile platforms also move into lower voltages to reduce power consumption like the new N series of Nokia phones that run on 1.8 V – sometimes this makes previous memory modules that run on 3.6 V incompatible with the mobile memory cards.

While processor capacity continues to follow Moore's law and double every 18 months, battery capacity has only increased 80 % in the past 10 years. This is unlikely to change even though there have been some hyped press releases every 6 months in the last couple of years about breakthroughs in the research of significantly improved battery capacity. So far the development has been evolutionary rather than revolutionary.

This has some important implications. In order for a hand-held device with limited battery capacity, to provide high customer satisfaction it needs to have well-integrated highly specialized power modules and chipsets. This means that technologies such as WLANs and DAB broadcasting that were initially not optimized for mobile use are complex to make user-friendly for small hand-held devices at a reasonable price. They require high industry investment that can only be justified by large volumes to reach an attractive price point. This 'chicken and egg' dilemma can only be overcome by joint efforts by industry consortia, which results in slower paced consensus decisions. The alternative is to grant a *de facto* monopoly, as is the case with Qualcomm's chip supply on the CDMA market. This has indeed resulted in quicker development and adoption of integrated network–device technology for operators (such as KDDi in Korea), but at the price of worldwide standards and global acceptance.

6.3.8 Memory and off-line synchronization

High-resolution color images and increasingly complex multimedia content (videos, 3D games, etc.) put increased requirements on memory capacity of the mobile device.

There are two types of memory in a mobile device: random access memory (RAM) and nonvolatile data storage. The first memory type is used for applications and the latter is used for static storage of programs or codecs – like a hard disk in the PC world or a memory card.

Memory capacity is steadily increasing and the cost is decreasing. For RAM, moving from SDRAMs to DRAMs implies a factor of 6 improvement in terms of transistors required per memory cell. The use of NAND memory instead of NOR memory reduces the cost per megabyte by a factor of 3. Memory capacity and easy-to-use synchronization technology are moving so fast that cheap offline material is becoming a cheaper surrogate for online downloads over expensive 3G networks.

In the Java MIDP standard, the Java Application Description, JAD, file contains information about the file downloaded. In addition to deciding whether the file can be downloaded (e.g. memory constrains), it is also used to check that the transmission was successful. The reality of today is that the so called 'jar-size' (Java archive file) is very limited. The most common gaming[12] mobile is Nokia's series 40 (series 60 has not gained so much market yet in the youth market due to its relative high price) is limited to 128 kb.

If we look into the crystal ball, we see that today's high-end devices with 2–4 Gb storage capacities (not available to execute the applications though) will be a commodity and follow the adoption rate of megapixel cameras. This implies that more than half of the sales of new devices will have this capability by the end of 2008. These devices will also have both USB and Bluetooth connectivity.

This is certainly good news for applications and media companies with mass-market ambitions of providing rich mobile content.

6.3.9 Display

The display itself is a key design criterion for applications. The multimedia experience is simply too rudimentary on some devices to be compelling enough to attract users. The impact on usage of having a color display is illustrated in Figure 6.8. It shows the number of page hits in an Austrian portal in Q42001–Q12002. During this period the Ericsson T68 was one of the few color devices in that market and, even though it had a low share of the total market, it contributed disproportionately to the service revenue share.

High-resolution color displays have become a key selling point. Numerous features and services make good use of them – GUIs, imaging, browsing and gaming. For the mobile device vendors this meant that in 2003 they could increase the average selling price and margins and sustain them in 2004.[13]

The display technology has evolved rapidly. In the last 5–6 years the pixel density has approximately doubled every 2 years; the graph in Figure 6.9 shows this development for mid-tier phones. High-end phones in 2005 have some 150–200 pixels per inch (PPI). As a comparison, the human eye has approximately a resolution limit of 700 PPI. Good printers exceed this resolution, which is one contributing factor to most people's preference to read printed papers.

The mass migration from monochrome to color screens has taken 2–3 years. The specialist research company Forward Concepts estimates that 85 % of the phones in 2006 will be shipped with color screens in the USA. This illustrates well how new device

---

[12] As much as 60–70 percent of the market in 2004 in some Western European countries according to Mforma.
[13] Forward Concepts (Carter L. Horney), 2005, 'Global cellular hand set and chipmarkets and Ericsson Mobile Platform'.

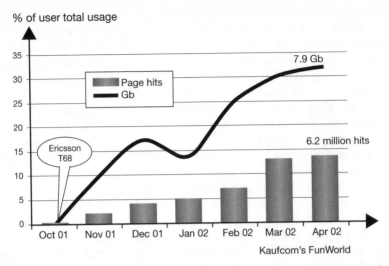

Figure 6.8 Illustration of the importance of superior displays, in this case using statistics of the number of page hits of the T68 (one of the few phones with color screens at this time) in an Austrian portal, Q12001–Q12002. Source: Ericsson.

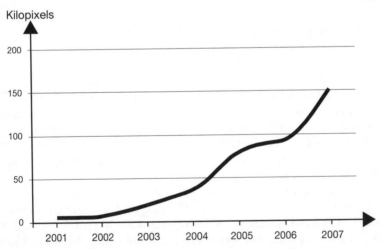

Figure 6.9 The fast development of display in mid-tier phones. Source: Ericsson 2005.

functionality diffuses from high-end segments to mass market, from being something for price-insensitive trendy yuppies and other early adopters in late 2001 to being a must-have for most mass-market mobile devices 4 years later, involving even late adopters. One can also see how the technology is diffused from the Japanese and Korean market to Europe and the US, reaching mass market 2–3 years later and at a slower adoption rate. The pattern is similar to the diffusion we see of camera phones.

6.4 OTHER IMPORTANT ENABLERS FOR APPLICATIONS AND MEDIA

6.4.1 Digital rights management

In order to create a lasting ecosystem for content delivery, it is fundamental to be able to protect the rights of the content owner and to make sure they receive royalties for usage. To make things more complex, there are different rights for the same content (performance, creation, etc.), but that is not detailed further here. While there exists proprietary sub-solutions for DRM, most operators are today focusing on the standards of OMA (open mobile alliance) for DRM. Much of this functionality has to reside in the device, but there is also a role for the operator. The main tracks are illustrated in Figure 6.10.

Forward lock means that you can download the content but you cannot forward it. This is the only mandatory part of the standard. With combined delivery a DRM message is created that contains both the content and the rights object. With separate delivery, on the other hand, content and rights objects are sent separately. This means that you can download a ringtone and the rights object with it. Then there is no problem in forwarding the ringtone to others, as they cannot play it without having paid for the rights object. This way of encouraging the spread of content but still ensuring that the rights owners get paid is called 'super distribution'. Super distribution and separate delivery are very important to mobile operators. Many content owners were not satisfied with the forward lock protection offered in DRM 1.0; they thought it was too easy to make pirate copies. The trend is to move from forward lock to super distribution. Super distribution features were included in the OMA DRM 2.0 Enabler. This was released in July 2004; with the typical lead-time this means that terminals should be available in abundance by 2005/2006.

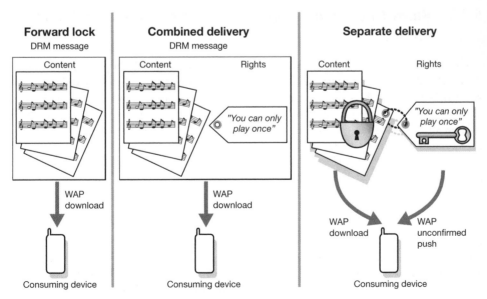

Figure 6.10 Different digital rights management methods.

6.4.2   IMS and SIP

When IMS has been introduced in both mobile devices and core networks over the next 1–2 years it will offer a common framework for introducing new multimedia services based on SIP. IMS is standardized in 3GPP, and its introduction is the first step in an overall network evolution to 'all IP', i.e. end-to-end (e2e) IP-based packet-only multimedia networks. IMS enables the operator to offer these multimedia services in compliance with traditional telecom characteristics, such as interoperability, security, and with high QoS. IMS also allows the operator to control what services are run and therefore also charge for IMS services.

The SIP, session initiation protocol, is an application-layer protocol that allows the creation and management of multimedia communication sessions between devices.

Instead of using proprietary protocols and interfaces for the applications, a developer can now base its applications on standard IMS protocols such as SIP. This will make it easier, for example, for a multiplayer game to work across networks and look similar in any operator's network, hence reducing complexity for the applications developer.

An SIP address (also known as a SIP URI) uniquely identifies a user regardless of the media that is used. The address resembles an e-mail address but has a 'sip:' prefix. For example, the fixed telephone on your desk might have the following SIP address: sip: caller@194.195.100.20. This will enable cross media applications, for example a video call between a fixed PC and a 3G phone, or a game session between a WLAN-connected P2P console and a mobile phone.

To give an indication of timing, the mobile platforms from Ericsson Mobile platforms will have OMA push-to-talk over cellular (PoC) and support for native SIP APIs for instant messaging and gaming in 2005, which means that there will be high-end devices in 2006.

6.4.3   Flash Lite – more consistent presentation layer for the mobile

Macromedia Flash is one of the most widespread applications with a penetration of more than 98 % of the world's market of 600 million PCs. It helps developers and designers to build richer content. It is mainly an enabler for the presentation layer of the application, as depicted in Figure 6.11. Apart from the user penetration, it has a large, more than one million strong, community of developers globally. Like content for web pages and GUIs for applications, Flash Lite brings interactive multimedia graphics to the mobile (and even the uses interface itself). This is compelling since a common goal of many application developers and media companies is to provide a harmonized, predictable user experience across all its channels, be it web, mobile or IP-TV.

Macromedia's Flash for mobile is an emerging technology for enhanced display capabilities. In Japan this has already started to take off: NTT Docomo markets its Flash Lite-enabled 900 series as 'fun-filled, colorful animations'. Nokia, Samsung and SonyEricsson among others have embraced the technology for their high-end devices initially. The system requirements of Flash Lite are fairly low: 1 MB ROM, 50 MIPS, 2 MB RAM and a 32-bit bus. This makes it technically feasible to embed it or download it to even low-tier phones over time.

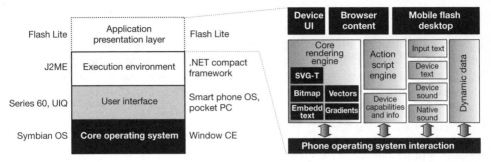

Figure 6.11 Mobile flash harmonizes the presentation layer across channels and devices.

Over time MacroMedia has the ambition to develop the script functionality of Flash Lite. It will then be more and more complementary to Java scripts, but focused on the GUI only. We believe the diffusion of Flash Lite will be large. It will follow a similar pattern to other new enablers that have received initial positive feedback. In 2004 it was first used in high-end phones in Japan; in 2005 it has commercial implementations in the European markets. By 2007 it will have spread to mid-tier phones.

6.4.4 Better browser standards

The device support of browser standards is essential to make it easier to adapt content for the mobile channel. With larger devices, the support for normal HTTP is also getting better through incorporation of a reduced set of the most commonly used Java scripts. Newer devices support WAP2, which handles the rich format of XHTML. In Chapter 4 we explored some of the challenges in developing browser-based application. It is hence clear that enhanced browser capabilities is one of the most important bridging enablers.

6.4.5 Over the air, device configurations and management

One of the fundamental aspects of the service environment is easy provisioning using OTA provisioning. This will be further explained in Chapter 7, but is definitely on our list of the key developments for the applications developer.

Nokia, Sony Ericsson and Siemens have a joint cooperation called Open Mobile Service Interface, allowing end-users to service and maintain content, software and the operating system of the mobile device. NTT Docomo have gone one step further to realize automatic upgrades over the air. These emerging standards are very important for the mobile applications industry since they enable continuous improvements and enhancements.

6.5 THE MOBILE DEVICE AS A DRIVER OF NEW APPLICATION FIELDS

The mobile device, independent of the development of the network services, is in itself a strong driver of market evolution. In fact the rapid evolution of mega-pixel camera capabilities, gigabyte storage capacity and high-resolution screens with the corresponding graphical processing capacity is surpassing the evolution pace of the wide area network capabilities. This has lead to a gap between the device capabilities and the network capabilities: customers will have mega-pixel cameras in their hands before the networks are upgraded to handle megabytes transfers economically and smoothly; thousands of MP3 songs can be stored today on the latest mobile devices but the infrastructure to download them is not ready in many parts of the Western world. Customers will instead upload their photos to their albums stored on PCs and transfer MP3s to their mobile music device; gamers will use WLAN-enabled games consoles to get the speed they need. This is of course a risk to mobile operators, but there is also a risk that the overall service adoption will be more device-centered and only provide good service in islands. This may not hinder early adopters from embracing the technology, but the mainstream (beyond the early adopter segment) wants things that work in an intuitive way everywhere. Hence, we see a large opportunity for the media industry and the applications developers to participate in driving and benefiting from the mass-market adoption of mobile services. We see the strongest clusters of mobile media and applications that have currently only penetrated the early adopter segment on a global basis around the following application fields:

- mobile music;
- mobile gaming;
- mobile TV;
- digital photo and video;
- mobile person-to-person interactivity (instant messaging, push-to-talk, photo sharing).

We will see more and more applications become 'mobilized' over time: digital cameras, game console, and video cameras. Figure 6.12 highlights this product extension.

6.5.1   Bridging enablers

Before we look into each of the emerging application fields we have identified, let us reflect for a minute over what the key enablers are that will bridge the experience gaps

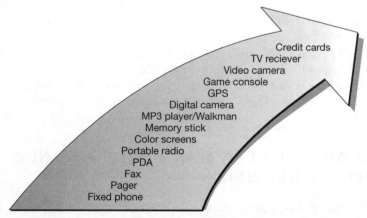

Figure 6.12 Over time mobility will be extended to more and more products and the mobile will be extended with additional product capabilities.

Table 6.2 Important bridging enablers

Gap	Area	Bridging enabler
Consistent presentation across web and mobile channel	Presentation layer	Flash Lite
Cross media and device interactivity	Convergent interactivity	SIP
Portable applications	Applications runtime environment	J2ME/MIDP2/+JSRs
Easy provisioning	Connectivity layer – access provisioning	Over-the-air activation
Economical delivery of mass market mobile TV	Better utilization of air interface	Evolved WCDMA (HSDPA, HSUPA)
Digital rights standards	Content delivery and protection	OMA DRM 2.0

between the mobile and other channels, because without a doubt there are still gaps. What needs to be in place to deliver a harmonized experience across different channel? What technology will help to adapt content to the mobile device and not just recreate but also enrich the experience?

Table 6.2 synthesizes the discussion we have had in this chapter an show the key enablers that can bridge some of the gaps we have to today in terms of inconsistent presentation, interactivity across different devices, easy provisioning and means of delivering mass market TV over the mobile channel.

6.5.2 Mobile music

Several of the dominant device manufactures have launched targeted mobile music devices. As we seen advances in storage, compression, battery life and wireless networks are making it easier to receive and store high-quality music on phones. Strategy Analytics estimates that in 2008 half of the 860 million cell phones sold will be able to store and play songs, up from 8 % in 2005.

What are the different vendors doing? In 2004 Nokia sold more than 10 million phones with integrated music players and 50 % of the shipments in 2005 are expected to have music functionality. Its multimedia 7710 smart phone features visual radio, which enables intuitive interactions with the radio stations: artist information, other essential data about the song, contest participation and purchasing of related songs and downloads. Motorola is leveraging its license agreement with Apple to use iTunes in its phones. This enables a similar interface to navigation and synching to iPod. SonyEricsson markets its mobile music players under the brand Walkman. Samsung has devices that leverage its internal hard-drive with extensive storage capabilities (>1 Gb), for example the SPH-V5400.

Even though in theory it was possible to use phones such as SonyEricsson's P800 when it was released back in 2002, it did not have good quality of sound, storage capacity or easy of use. For the mobile phone to be an attractive music player, a whole range of criteria need to be fulfilled: storage capacity, long battery life, stereo sound with high voice quality, headphones, sound chips and procession power, simple synchronization and intuitive user interface. One should not expect the music-phone to be able to compete with specialist iPod devices. If you compare SonyEricsson's over-the-air music download service PlayNow and its music-featured phones in the Walkman series to online downloading over broadband and the 60 Gb iPod, one realizes that the two serve different purposes. The first is more a 'seven–eleven shop' with a limited top-pick selection and the second compares to a large grocery store where you go and do your week-end shopping and buy all you need.

Today, 'wireless' music stores cannot compete with lower-cost, better GUI dedicated solutions over the fixed Internet and specific device. The key is mobility and interactivity bundles (alerts, video and exclusive artist information). This said, it is interesting to compare the uptake between KDDI's music services Chaku-uta (clips of 30 s), Chaku-uta-full (full songs over the 3G network) and i-Mode (Figure 6.13).

6.5.3   Mobile gaming

Gaming and betting is a multibillion dollar business and the mobile device is the perfect media with its true mass-market penetration. Downloaded games require a execution environment like Java or BREW. An important amendment to the Java standard for

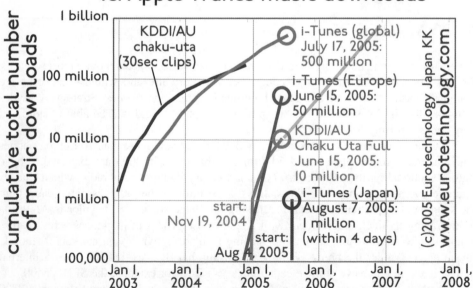

Figure 6.13 Comparison between Japanese music offerings and i-Mode. Source: Eurotechnology, 'KDDI's success story', 2005.

games is the Java Standard Community (JSR) 184 that adds console-quality 3D graphics to mobile devices.

Nokia's N-gage phone is one of the first examples of a phone designed with gaming in mind. This platform also offers an alternative physical distribution of games in the form of MMC memory sticks. As mobile Java technology becomes integrated into low-end devices as well, gaming will reach higher adoption levels. One of the challenges today is that the target market of youths 15–25 years old do not have the devices that enable compelling games. Series 40 of Nokia is still the dominating device with its limited jar-size. In the coming years we will also see a trend for games consoles to have connectivity (for example Sony's portable P2P games console) to a larger extent, in particular through WLAN-enabled cards.

Case study: Jadestone developing cross device and cross media games[14]

Jadestone is an innovative game developer and the second largest gaming studio in Sweden. They have developed several successful online games for Internet-enabled PCs. Jadestone was also the first to launch a multiplayer game using Nokia's N-gage terminal, with titles like Spirits and Age of Piracy. Their current online games are now also being expanded with a mobile client. Jadestone see the following benefits with the client-server model they have chosen:

- a persistent game world, always on, always with you – wherever you are;
- instant synchronization between mobile and PC clients;
- community, friends lists, forums, clans, presence info;
- optimized communication protocols for real-time multiplayer gaming;
- longevity of game play;
- 100 % piracy protection.

Around the corner they see further cross-platform integration between PC and games consoles but also tighter collaboration with brand and content owners such as Eidos. In Jadestones's view, it is not about recreating the experience identically across different platforms. It is instead important to create a complementary experience that uses the specific traits and strengths of each channel. In Jadestone's experience, the mobile channel is good for community building and maintaining contacts with people you already know, but less attractive for exploring new contacts and objects. So, what are the challenges in going from the web channel to the mobile channel?

Lika Mforma and Terraplay (see previous case studies), Jadestone see the device fragmentation as a big challenge. They have to date not used specific porting companies to deal with the porting issue, like Mforma. Jadestone claims that the porting companies can do porting of client games well but have limited experience in the online client-server games Jadestone develops. Instead they have developed their own framework, 'Forward', to deal with porting as early as the design phase. The framework is made up of two parts, the build-time generation of user interface data and the run-time interpretation of the data and creation of actual interface controls. Using this concept, designers take over

---

[14] Source: Jadestone.

the responsibility of positioning GUI components in a screen and connecting them with navigation links. The basic idea behind this is that most views in the GUI of the mobile applications are rather simple, almost always consisting of standard components with very little or no view-specific behavior. This facts, combined with the experience that developers of mobile applications spend a lot of time rearranging GUI components in the views and redefining the navigation between them instead of writing game logic code, led to the idea of Forward.

In Forward, each view (or control set) is described in an XML format in a control set description (CSD) file containing information on the view and its controls. A control set could be a text label, an image or an input field. The description of components includes their characteristics, such as the texts of labels or the image files to use for picture controls, as well as their position in the view and the navigation between them.

At build time the information in the CSD files is used to create binary encodings of the view: Forward (FWD) files. The content of the FWD file is the same as that of the CSD file, but the format is significantly more compact. At runtime the view is constructed by reading the FWD file. When reading the FWD file, the controls specified in it will be created and tied together according to the navigation described in the CSD file.

When porting applications the work is done towards predefined groups of phones from one or more manufacturers. Phones are grouped according to characteristics such as screen size, MIDP version, heap size and maximum JAR size. The aim is that a game client built for a specific group of phones should run on all phones within the category without modifications to the code.

6.5.4 Mobile TV

One of the most successful nonvoice services to date among 3G users is mobile TV. Penetration of mobile TV has exceeded 30 % in South Korea and, more than 50 % of its 3G customers have adopted the service.[15] For Western Europe it is expected to reach 10 % penetration in 2007. Vodafone supports this conclusion and predicts a significant spread of the TV on mobile service until the end of 2006/2007. Its research concludes that there definitely is a demand for TV on mobile, general and local news being the most desired application, followed by existing TV programming and short movies/trailers.

In Korea the rapid take-up has led to poorer quality of service, which has had a negative impact on the service perception. This is because the current networks are not efficient in delivering broadcasted content, which translates into a high cost in terms of spectrum usage.

Commercial mobile TV services in current mobile networks are being offered through streaming and Unicast solutions. These solutions may face capacity problems if take-up is as quick as it was in Korea. That will be partly solved with the introduction of HSDPA. On the other hand, these solutions do not allow the delivery of 'multicast' services until MBMS (Multimedia Broadcast and Multicast Service) functionality is included in 3G networks, which, according to 3GPP, is planned for commercial service by 2006.

---

[15] Analysys 2004 (Alex Zadvorny), 'Making a success of the mobile content value chain'.

MBMS (using 3GPP terminology) or BMCS (Broadcast and Multicast Services using 3GPP2 terminology for WCDMA 1xEV-DO) is the proposed standard for delivering live video via the networks of mobile operators. With MBMS implemented, the radio transmission cost is independent of the number of users in the cell. It is particularly efficient when there are many users in the same cell of radio coverage. However, this is not likely to replace the use of Unicast for customized content. As on the Internet, Unicast is becoming more and more often the prime choice.

Another 'competing' standard is digital video broadcasting (DVB-H). DVB-H is a version of the existing DVB-T terrestrial digital TV standard but modified to suit mobile, battery-powered terminals. Rather than broadcasting continuously, DVB-H broadcasts in bursts, allowing the receiver to power-down whenever possible, boosting battery life. However it requires a new access network and spectrum allocation is not resolved on a global basis. Another issue is the poor indoor coverage.

DVB-H trials are underway in the UK, USA, Germany and Finland, with a view to rolling out services some time in 2006. Nokia is working on a DVB-H device with a view to a 2006 debut. On the application layer, Java will be an important means to run applications that interact with TV through mobile devices or web clients. The BBC among other broadcasters are running trials to evaluate this further.

6.5.5 Digital photo and video

We have talked about the technological and market diffusion aspects of the camera mobiles, but what is required to create attractive services around the device? As we saw, to date digital photo and video have only reached the early adopters on a global level. To drive adoption beyond the early adopters, things most be simplified. Today many users are disappointed with both the poor quality and the complexity of their camera phones. Ease of use becomes a means of differentiation for suppliers and application developers. Over the next couple of years multi-mega-pixel cameras will hit the market, which means that the technology will be in the hands of the consumers that will enable them to make printouts with high resolution and store high-quality digital albums. An ecosystem around the camera phones will develop:

- memory cards to transfer data to different devices (PC, TV, mobile, printer);
- personal area network connectivity (USB card, WLAN, Bluetooth);
- printers, in the home, at the office or at printing shops;
- community sharing through blogs, etc.

This will give rise to many different applications to knit all the entities together and simplify the management, sharing and distribution of photos.

6.5.6 Mobile person to person interactivity

Today, we usually choose first how we wish to communicate, such as by selecting SMS or MMS, then define to whom the message should be sent. The introduction of IMS enablers such as a presence-enhanced phonebook (PEP) into the terminal as the starting

| Who | How | Add participants
as you go | Add media
as you go |

Figure 6.14 Presence-enhanced phonebook and other IMS features enable new ways of starting and expanding communication.

point for communication will change the way we communicate. We can select the most appropriate communication type depending on the status and mood of our friends and colleagues. By being able to shift between different means of communication, the user has a more flexible and personalized way to communicate (Figure 6.14).

New communication services that can be foreseen today are:

- presence-enhanced phonebook, including group list management;
- push-to-talk and push-to-video;
- integrated communication with instant content sharing (started as a combinational services work item in 3GPP);
- instant file/URL transfer;
- peer-to-peer gaming;
- instant messaging/chat;
- one-to-one or one-to-many voice/video;
- whiteboard conferencing.

Our view on IMS clients in the mobile device is that there will be both embedded and downloadable IMS client applications. Critical factors from a device perspective in this will be: easy and rapid development on clearly defined APIs, coexistence of embedded and downloaded IMS clients, smooth downloading of new IMS clients.

The availability and interoperability of clients and terminals supporting IMS-based services is the key for future success and rapid end-user adoption. Even though this will happen as soon as 2005/2006 for high-end phones, it will take 2–4 years before it is available globally, including the roaming capabilities. Note that it is not technically feasible today simply because it takes time for operators to implement it and for application developers to rely on it.

6.6   MOBILE DEVICE ADAPTATIONS

The Japanese operators, in particular NTT Docomo and KDDI, showed the importance of driving the evolution of the device capabilities and configuration through close cooperation with the device manufactures. Vodafone, through it consumer proposition Vodafone Live, showed that it was possible to replicate this in markets outside Japan. The 'Oscar' of mobile communication, the Cannes award, was given in both 2003 and 2005 to operator-specific devices developed in close cooperation with the Live! Program of Vodafone, the Sharp GX10 and SonyEricsson's V800. Nokia was initially hesitant to produce device adaptations, but has now embraced them with its new platforms being configurable in terms of both hardware and software.

Vodafone found a clear correlation between service revenue and the degree to which the device was customized. When Vodafone Live was launched in 2002, there were three devices: the Sharp GX-10, Panasonic GD87 and Nokia 7650. The most customized and optimized device for Vodafone Live was the Sharp GX-10; it also showed the highest service revenue, as indicated in Figure 6.15 (note that the numbers are only illustrative,

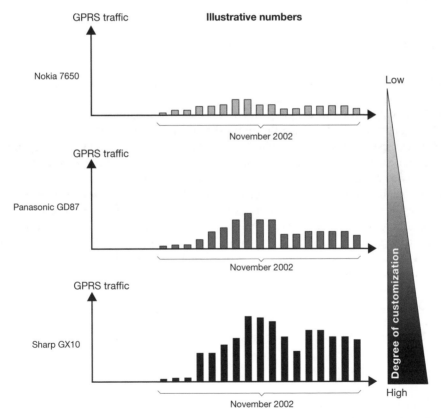

Figure 6.15 Difference in service revenue based on degree of customization. Source: absolute numbers in this picture are only illustrative, but are based on observations of usage the first month of Vodafone Live, November 2002.

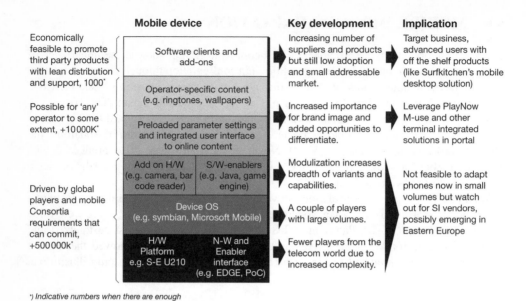

*) Indicative numbers when there are enough
economies of scale, as of October 2004

Figure 6.16 Adaptations of different layers of the mobile device. SI, system integration.

even though they are based on actual observations). There was a clear correlation between usage and degree of customization.

In the mobile device business, economies of scale are extremely important. A player like Vodafone can afford to drive the entire specification of the mobile device and makes changes and specifications with implications for both the hardware and low-level software layers. The Sharp GX-10 sold more than one million devices. To change and design a mobile device on the lower level, several hundred thousand devices are generally required to make it economical feasible. The increased modularization of the devices in recent years has continued to push down the volumes required to break even. The modules become more and more open but at the same time more complex. That makes it necessary to allocate the development cost of the platforms to larger volumes. If one uses a standard device platform but makes a specific integration of add-on hardware like additional memory or a camera, then volumes of 100 000 could be feasible depending on the complexity of the integration. Preloaded parameter settings and specific branding elements like ringtones, wallpaper or games require much lower volumes, in the region of 10 000. Figure 6.16 summarizes the changes at different levels of the device, indicative volumes required and some key trends and implications of this.

6.7   SUMMARY

We have reviewed the impacts of the rapid pace of development of the key components under the hood of the mobile device: higher processing capacity, larger color screens, higher resolution cameras and more storage. To this we have tied the important development of software enablers, in particular Java. What we see is that most components are already in place for very powerful applications in the high-end phones to be released now. However, with the power of the >700 million units-a-year industry that the mobile device business represents, we see a rapid diffusion of the technology to affordable low-end phones. The economies of scale and the fierce price competition make the mobile a truly interactive and attractive channel for media and content for the mass-market in the near future.

7

Service Environment

7.1 UNDERSTANDING THE SERVICE ENVIRONMENT

It is becoming clear that the basic need for communication between people is, and will in the foreseeable future remain, the dominant driver for the telecommunications business. It is also clear, however, that the role that telecommunications play in the lives of users is expanding beyond voice to new services that allow people to do more than just speak to each other, things like utilizing sound, text and video in their interactions. Users want to access media, information systems and services from any place and at any time, and gain greater control of over how they interact with the outside world.

The term 'service environment' is used to refer to the entire business and technological context in which mobile services are successfully, efficiently, securely, reliably and profitably traded between the users and providers of those services. This seemingly abstract term is deliberately chosen for a few simple reasons: firstly we would like to avoid confusion with existing terms involving the use of words like *platform, system, network*, etc., as we believe that successful service delivery is about much more than technology. Secondly, we would like to avoid cornering this area into any one discipline, technology, standard or actor; we believe, simply, that all of these aspects must unite before the true promise of mobile services is realized.

At this point, we would like to make an early attempt to visualize the service environment given what we have stated so far.

On the edges of the service environment are the people and enterprises that use services for business and pleasure. They call each other, send messages to one another, browse the web, play games, download music and so on. It is the need of the people and enterprises at the edges to interact with each other for various reasons that generates business, and the satisfaction of those needs at competitive prices that sustains business.

Beyond the edges exists the infrastructure that supports the delivery of those services. This infrastructure stretches from the device and its software to the access, core and backbone networks, to business support systems and service enablers, to the Internet and

Mobile Media and Applications – From Concept to Cash: Successful Service Creation and Launch   Christoffer Andersson, Daniel Freeman, Ian James, Andy Johnston, Staffan Ljung
© 2006 John Wiley & Sons, Ltd

Figure 7.1 The service environment is about people.

the networks and infrastructure of content, service providers and the devices of other users.

In short, the service environment is characterized by a diversity of actors, technologies and ambitions and is a true end-to-end technological and business environment. Moving onwards, we would like to discuss the service environment anatomy as a kind of *service trading marketplace* and *machine* duality.

Our aim here, is not to present a *grand unified theory* of mobile service environments but rather to draw attention to the interdependence of business and machinery and even more importantly to emphasize the necessity that these two essential components evolve in mutually reinforcing ways with user value and business viability in focus at all times, a necessity which we feel we have often lost track of in the brief history of our industry.

Using our anatomy, the outer shell of the service environment represents a *service-trading marketplace* with its services, actors and business models. This outer shell, in turn, wraps an inner core of machinery and technology supporting the required functions and expected qualities of the marketplace. As with any living and evolving organism, the boundaries between shell and innards, between crust and core, are not entirely clear and are constantly shifting – our service environment anatomy neither is, nor needs to be, an exact science.

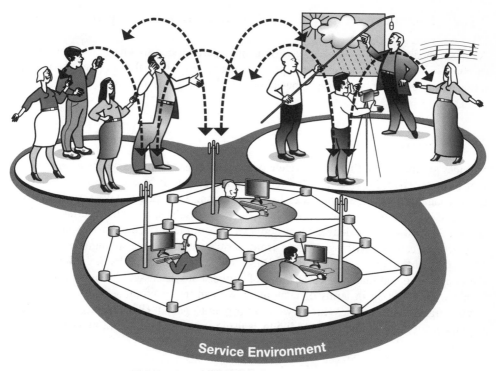

Figure 7.2 The service environments is about business and technology.

Figure 7.3 The service environment is both a marketplace and a machine.

7.1.1   The service environment as a service trading marketplace

As mentioned above, the service environment is part service trading market place, part machine. The service trading marketplace is characterized by the services themselves, the actors and players and by the business models and value chains that connect these participants.

7.1.1.1   Services

Services are the reason that the service environment exists from the business and technological perspective. Services are the items that are produced by some, offered by others and consumed by yet others. Services are traded amongst all parties in the service environment and it is the features and appeal of services themselves, as well as certain qualities in, and the cost of, their delivery that defines *value* in the service environment.

A service can be anything ranging from basic connectivity to a ringtone download, MMS postcard, a Sports Score update, SMS voting and so on. The list is limited only by the imagination and many different kinds of services are discussed throughout this book. While we are talking in terms of anatomies, we can try to group services into the following high-level categories: firstly we have person-to-person (P2P) services that include communication services between individuals using voice, text or picture messaging, video

Figure 7.4 The service environment is about trading.

telephony, push-to-talk and so on, secondly we have person-to-content (P2C/C2P) services providing access to content and media, for example browsing, downloading or streaming of news, sports, music, pictures and so on. Last but not least, we have *enterprise services* providing corporate users with access to enterprise data and to productivity applications such as e-mail, calendar Internet access and so on.

Naturally, new service categories will be established as new ideas are born and technologies arrive to support them. Conversely, new technologies will be invented in order to support new service ideas. This symbiotic development leads to one of the trickiest aspects of service environment development and evolution, a kind of *technology-chicken* and *application-egg* problem. More often than not, a technology exists before the services that use it are in place, the development of that technology subsequently suffers from the lack of real-world use that would test, tune and direct its development in the best direction. This leads either to long inefficient paths to satisfying real-world applications or to disappointing first uses of a technology that may blunt both its initial impact and future prospects on the market. In recent years, the industry has taken an increasing grip of this problem and made a conscious effort to work on concrete applications alongside development of platforms and technologies, yielding a more pragmatic overall evolution of technologies.

The case of combinational services presented below represents how an appreciation of the necessity to combine technological advances with user-satisfying and business-viable applications of those technologies has led to changes in the way innovation proceeds and the way in which existing technologies can be harnessed in creative ways to deliver new services ahead of the technology curve.

Case study: combinational services – a pragmatic first step to All-IP

The idea with combinational services is to take a pragmatic first step towards services foreseen in the All-IP world, but before All-IP becomes a reality.

Today, our industry has two main deployed technologies at its disposal; these are circuit-switched mobile voice networks and packet-switched access in the form of GPRS and other technologies. The 2.5G radio-borne packet-switched network currently has some

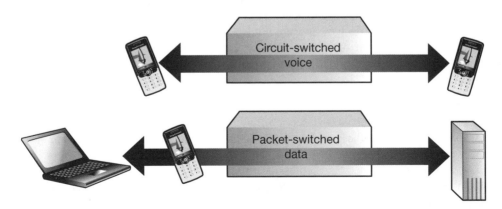

Figure 7.5 Circuit/packet switched services.

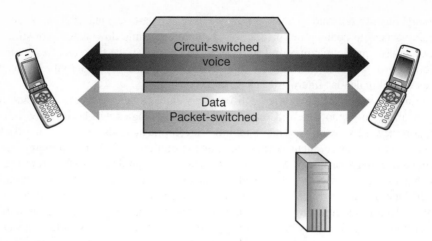

Figure 7.6 Combinational services (using circuit/packet switched).

limitations when it comes to implementing voice over packet. Chief among these are longer handover delays, not noticeable when using the web, but significant for voice conversations; and additionally, the asymmetric nature, with lower bandwidth/higher latency in the uplink that becomes a limiting factor when applied to an upstream of voice from a device.

As these limitations will be with us for some time, the notion of *combinational services* has been coined to refer to an interim way of delivering enhanced services using the existing properties of both circuit- and packet-switched networks simultaneously – real-time parts of the service are delivered over the circuit switched network while nonreal-time parts are delivered over the packet-switched network.

An example of a combinational service is a situation where two conversing people share a picture with each other in near real-time while chatting *on the phone*; the circuit-switched voice call continues while a packet-switch-based picture transfer proceeds simultaneously.

A number of, evolutionary rather than revolutionary, technological advances are necessary in order to make combinational services possible; firstly multi-RAB (Radio Access Bearer); support capability is needed in the radio access networks; secondly dual transfer mode support, allowing the simultaneous use of both packet- and circuit-switched services, is required in the devices; last but not least, reachability and capability negotiation is required end-to-end in order to allow user equipment and intermediate networks to agree on the feature set they can support together.

7.1.1.2 Actors

Beyond services themselves, the service trading marketplace is home to a number of business actors or roles; these are the people and organizations whose needs and ambitions power the service-trading marketplace.

Users such as consumers and professional users are the viewers, listeners and users of content and services as well as the users of communication services. Users are arguably the key players in the service environment – it is their needs that fire the service trading marketplace and their satisfaction that sustains it.

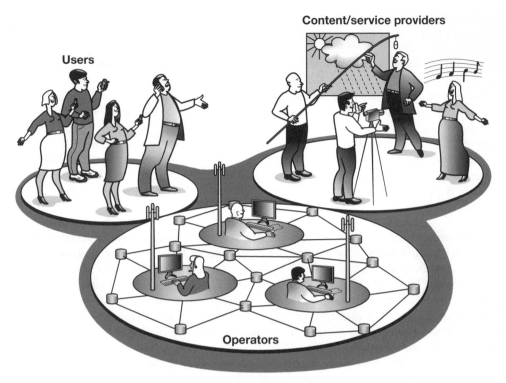

Figure 7.7 Actors in the service environment.

Operators provide machinery supporting service discovery, management and delivery. Typically, this equipment ranges from the devices and radio network to switching for circuit- and packet-switched services as well as enablers for services such as download, streaming, messaging and so on. Operators also run systems supporting key business processes such as billing, provisioning and service assurance. The operators have a special role as key investors and drivers in service environment innovation; they also occupy a position as first point of contact with the user and therefore play a vital role in overall user satisfaction.

Content and service providers may be either independent businesses or a role that is played by another actor such as an operator. Content and service providers both create and supply content and services to the mobile service environment market. Creation and production of content and applications is done in their own environment with operator network channels facilitating delivery, optimization, security and so on. Content and service providers are responsible for and generally have a customer relationship with their users for the services and content they provide. The independent content and service providers are key entrepreneurs and service innovators in the marketplace, adding and keeping energy and life in the marketplace.

The nature of business means that actors, especially continuously thriving actors, have to change shape over time. In some cases, a successful actor may see an opportunity to expand upon their existing responsibilities; in other cases, technology developments or an evolution of user or marketplace needs may mandate that an actor evolve in one or the other direction or may yield opportunities for new actors to emerge.

Case study: actor evolution, operator as an identity and trust provider
In recent years, developments along a number of axes, have, along with the operators existing role and technical capabilities, somewhat happily aligned to present the mobile operator actor with what some consider an opportunity to assume a new role in the overall service environment – that of the *identity provider*. The forces leading to this opportunity are manifold: firstly there is a heightened awareness of the need for online security; secondly, the mobile services marketplace is expanding beyond the walls of the operator to a community of content and service providing entrepreneurs; thirdly, standards bodies and companies are providing interoperable technologies that make all of the above possible.

An operator that takes on the role of identity provider can assert the authentication of a user towards other actors. When a user is attached to the operator network via the secure mechanisms of that network, the operator can *vouch* for that user's *identity* online towards other actors systems. The Liberty Alliance Project speaks in terms of communities of actors in a service environment forming *federations* that form circles of trust amongst each other; technologies openly specified by the Liberty Alliance Project (including those for identity providers) can be used to technically facilitate the sharing of identity (and more) across such federations.

The existing business role of the mobile operator, which includes a contractual relationship with the user and implies a strong level of trust, combined with the operator's technical capabilities, which include secure mobile network-based identification and authentication of the user, provide fertile prerequisites for operator evolution into an identity provider role.

The notion of federations of operators and service/content providers has established itself conceptually and practically in commercial contexts. Operators are acting as identity (and billing/payment) providers on behalf of their subscriber base towards service providers and in certain markets this is already the dominant business model. Currently, however, no single technological approach has gained a dominant position in this area – yet.

7.1.1.3   Business models

The actors in the service environment maintain relationships with each other in accordance with business models; a business model is simply a way of describing the ways in which actors do business with each other, including responsibilities, value chains and so on.

Operator as service provider: in this business model, the operator purchases content and/or applications from service/content providers and resells these to users. The operator takes full responsibility for the function and quality of the services and content towards their subscribers and is their primary technical and commercial point of contact in the service environment. In this model, the operator's service environment machinery houses much of the service logic and content forming the backbone of the service offering.

Initially, having such a rich environment was seen to be an advantage in terms of possessing a complete offering – in recent times, it has become evident that it is difficult

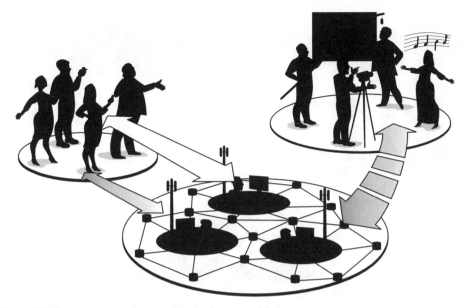

Figure 7.8 Operator as service provider business model.

to achieve the agility required by the market in a service environment that attempts to do too much.

Operator as channel provider: in this business model, content and service providers sell their content and/or applications directly to the operator's subscribers. The operator provides a channel for these sales; the channel includes delivery mechanisms such as the basic wireless packet connectivity, messaging services and so on. The same channel may include line-of-business supporting machinery such as those for identity management, payment, service assurance and digital rights management, as well as service-enhancing user-related information such as device characteristics, user's location and so on. The subscriber generally has a relationship with both the operator and the content/service providers whose services they use. The operator may also offer content/service provider *hospitality* features such as self-service portals, service creation toolsets, test mechanisms, etc.

The channel provider business is already viewed as a viable volume business for operators. A large number of mobile operators view opening their minds, organizations and service environment machinery to content and service providing entrepreneurs as one of their biggest challenges and a key factor in their business success going forward.

Operator as bit pipe provider: in this model, the operator is focused on the task of providing wide area wireless IP bit pipe services to users (or to any actor having a need for such access). It may be contended that each and every operator needs to be at least an excellent wireless bit pipe provider; however, some operators may chose to focus solely on that proposition and strive for the technical and/or commercial high ground in that area by having the best coverage, best capacity, lowest latency, highest availability, differentiated service offerings, leading wireless IP optimization techniques and so on. IP is the common denominator going forward and being an excellent wide-area wireless IP provider, though certainly not trivial, has an immediate and proven business case.

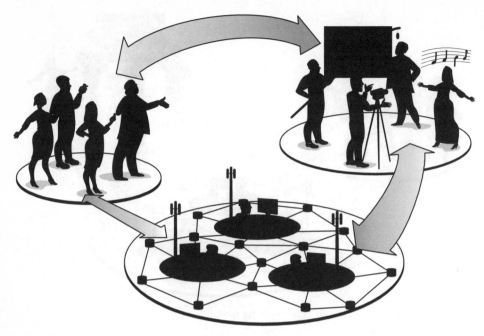

Figure 7.9 Operator as channel provider business model.

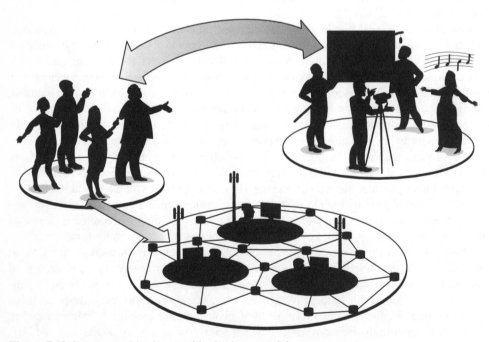

Figure 7.10 Operator as bit pipe provider business model.

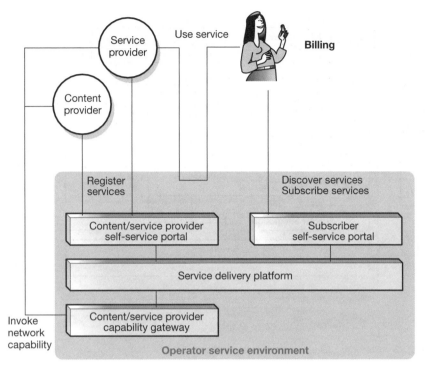

Figure 7.11 Aspiring 3G operator bets on channel provider model.

Case study: 3G greenfielder bets on channel provider business model
In this section we take a brief look at how an emerging 3G operator is staking their future on a business model, predicated on a channel provider model.

This operator foresees that the majority of services, with the exception of person-to-person services, are to be provided by a network of service and content provider partners who register themselves and their services with the operator using the specially developed service and content provider self-service portal.

Once registered, the content/service provider is granted access to the content/service provider capability gateway (for example a Parlay gateway) and their registered services are available for discovery and subscription by the operator's subscribers at the subscriber self-service portal. Once a user subscribes to a service, they can personalize it at the content/service provider's own web/WAP site. The operator acts as a connector between all players, providing authentication and single-sign-on capability, service authorization, charging, discounting and payment settlement services, while allowing for the existence of a thriving, innovative and dynamic content and service provider community.

7.1.2 The service environment as a machine

In the last chapter we looked at the service environment as a service trading marketplace, in this chapter we will introduce parts of the *engine room* for this marketplace; the

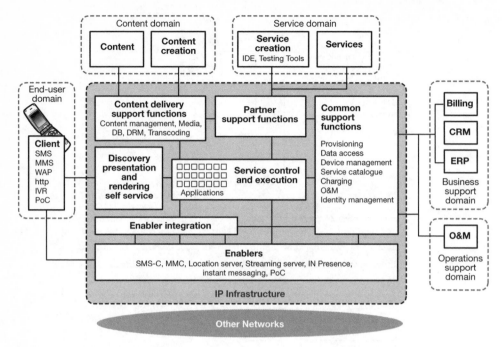

Figure 7.12 Inside the service environment machine.

functions, processes and technologies involved in producing, delivering and managing all aspects of the marketplace.

Operators sometimes use the metaphor of an operating system in referring to their service environment machinery; like a real OS, it offers a programming model for services, it also provides an execution model and environment for code that implements those services, and, similar to a real OS, it takes on the task of steering a number of peripherals like the operator's service enablers.

Figure 7.12 visualizes, albeit on a high level, the anatomy of a service environment machine. In advance of taking a look at each of the constituent parts it is perhaps interesting to note that the shape of Figure 7.12 has really settled down in recent years to the extent where many in the industry might now readily find themselves in a position to agree broadly with the functional divisions and responsibilities outlined in it, although opinions will naturally differ on how to best implement and achieve those responsibilities.

7.1.2.1 Devices and clients

The user's device is one of the most vital parts of the service environment machinery and a complex piece of machinery in itself. It is so important that we devote an earlier chapter of this book to it and therefore we will not heavily elaborate on device-related aspects in this chapter. It is, however, worth spending a few moments on the changing nature of the relationship between the devices and the operator service environment; in the early days of mobile services, device and operator service environment capabilities marched ahead together in lockstep, guaranteeing function but yielding a rather slow evolution dictated by the speed at which networks could be evolved.

Today's reality is a different one; the pace of device development is less strictly coupled to operator service environment evolution, leading to a more dynamic overall evolution of both. This dynamic also brings with it a number of challenges. Chief amongst those is the fact that users legitimately expect that things work in as simple a manner as possible, which in turn places new demands on the service environment to support seamless integration and management of a wide array of continuously evolving devices, a topic we touch on later.

7.1.2.2   Content creation, management, delivery

This part of the service environment machine room is concerned with the entire lifecycle (creation, ingestion, transcoding, tagging, editorial control, digital rights management and so on) of the content that we hope will pulse through a lively service environment. It contains things in the operator environment, on the content creator's desk, in the content provider's environment and interfaces and formats between all of these. We take a deeper look at some aspects of the content management machinery below.

7.1.2.3   Partner support functions

This is the area of the operator service environment where entrepreneurial service and content providers are connected in, serviced and enabled. As with content management, this part of the machine room needs to support the entire lifecycle of the partner (self-service, service creation, service publish, service subscription notifications, network service invocation, single sign on, user and network attribute exchange, service level agreement management and service assurance). Partner support function machinery is examined further below.

7.1.2.4   Discovery, presentation, rendering and self-service

The browser-dominated access model places a special emphasis on the place in the service environment where the service offering is exposed to the users – a place commonly referred to as a *portal*. The portal parts of the service environment machine support service discovery and subscription as well as personalized user access to those services. Portals are the operator's (or content/service providers) shop window and front doors at the same time. We will take a look under the hood of the portal a little later in this chapter.

7.1.2.5   Enablers and enabler integration

Enablers are key pieces of machinery, providing functionality and interfaces towards *lower* network layers in the operator service environment. Enablers allow systems in the service environment (such as portals and devices) to access the services offered by underlying layers in the circuit- and packet-switched core network and other networks.

Enablers are often put in place as a result of the development of *standardized* services such as WAP, MMS and streaming. Enabler integration refers to a set of generally operator service environment architecture-specific middleware used to harmonize integration of a diverse set of enabler management interfaces into the common support functions that are part of a horizontally architectured operator service environment. Later we look at the collection of enablers that today constitute a typical service environment.

7.1.2.6 Common support functions

Historically, operator service environments have grown in a rapid fashion under time-to-market pressure by the continuous addition of vertically integrated applications and services. Recently however, there has been a clear trend towards horizontally structured operator service environments in which commonly reused functionality, most often supporting key operator business processes, is separated out into individual systems and made available on open stable interfaces and mandated for use by all other concerned systems. Examples of such functions are those for: user profile, provisioning/activation, charging, operations and maintenance, identity management, user/partner authentication, user authorization and session management. The topic of common functions and horizontalization is something we will return to in a later chapter.

7.1.2.7 Service control and execution

The service control and execution area deals with the ways in which the application code that forms the heart of services within the operator service environment is executed and how that execution is controlled or coordinated with events in the network. In recent years, there has been a clear trend away from models where the majority of applications execute within the operator service environment to more devolved models where applications execute in partner environments, out on the Internet, in devices or across all of these. There is still, however, a need to deploy and execute certain classes of application within the operator service environment; examples of such services are intelligent network (IN) services such as number translation or follow-me services, voice-centered services such as personal assistant, services in the mobile network such as mobile positioning, as well as services emerging in All-IP networks using IMS technologies.

A number of software technologies are at hand to enable service control and execution for these kinds of services. One interesting development in this area is service logic execution environment (SLEE) technology. The aim of SLEE is to provide a common programming model (API) and a run-time environment for the development and execution of Java-based network services. The SLEE development model is conceptually similar to that available in Java-based enterprise computing environments today; however, the implementation is tailored to the fine-grained event-driven nature of telecommunications applications. An ultimate goal of SLEE is to lead to portable interoperable service implementations that can be deployed on any SLEE compliant environment, much like what is possible (with some care) for J2EE applications in the enterprise space today. Another obvious goal is to reduce development effort and complexity.

7.1.3 The battle for mindshare

The evolving openness of the service environment has led to the entrance of new players and ideas and the service environment has become battleground for commercial (and

Figure 7.13 Interests and expertise of all kinds converge on the service environment.

conceptual) supremacy. This increased openness bodes for optimism that a greater number of new interesting services will emerge as network capabilities become available to a wider, more innovative and more entrepreneurial community. The higher levels of competition, not only among vendors and system integrators but also amongst operators, service and content providers, should lead to better value for consumers and also a higher benchmark for the quality of overall service delivered to the users and business whose pockets form the financial backbone of the entire business ecosystem.

The evolving openness of the environment also brings with it a *competition of concepts* where commercial interests sensing opportunity begin to ramp up their marketing and concept machinery and attempt to create or gain market share by putting a new spin on established problems and solutions or by injecting new or previously un-leveraged technologies into the environment. This dynamic has both positive and negative sides to it.

On the positive side, new concepts and technologies can have transforming power when applied to an area, and in some cases, certain established practices have long passed their sell by date and are an obstacle to progress. On the negative side, the competition of concepts sometimes creates an over-supply of jargon and claims of benefits that sometimes make it hard to cut through the claims and conceptual blanketing to the essence of the problem definition and solution offering.

7.1.4   Further models for the service environment

In this section we look at some models from the technology arena and beyond that have something to offer in understanding the service environment.

7.1.4.1   The service environment as a department store

The department store model has existed for hundreds of years and has been tried, tested, tuned and improved over its lifetime. Department stores aim to provide a single place where many needs can be fulfilled and offer more convenience than going to several stores; they aim to offer a reduced purchasing risk and a more consistent satisfaction guarantee. The department store allows the shopper to focus on their needs and the products, not on aspects of how to find them or whether they can trust the vendor.

The department store model translates not only to the business acumen and commercial sphere but also into the design, planning and architectural spheres. To illustrate this, a colleague drew the analogy with a supermarket: 'Imagine your local supermarket changed the location (portal design) of its entrance every week, unsystematically changed its shelving (service discovery design) or had to exchange its cash registers (common function design) every time a new even-better brand-x soap powder was launched'; hence, the service environment designer can and probably should try to learn a trick or two from the department store or supermarket designer.

7.1.4.2   The service delivery platform model

Taking a looking at the functions and qualities available in service delivery platform (SDP) offerings today offers many helpful insights into the needs for operator mobile service environments. SDP are business solutions aimed at the mobile operator segment. SDP generally combine common functions, communication infrastructure and mobile service enablers as well as facilities for creating and exposing service capabilities to service and content providers. SDP aim to provide a systematic, integrated way of dealing with an operator's service environment capabilities.

The stated aim of the SDP is to enable faster implementation cycles and easier system integration efforts, yielding a technically and economically sustainable evolution of the operator's service environment. A general goal of the SDP is to provide reusable base components that have proved useful in supporting typical operator needs. Functionality such as single sign-on or provisioning as well as new categories of services can be implemented step-by-step according to the operator's business strategy. Service evolution becomes more predictable, making launch and marketing programs easier to plan and coordinate. In addition, SDP are designed with key qualities, including scalability, reliability, security and manageability, in mind. As operators often have a portion of the SDP functions already installed in their existing networks, SDP are generally flexible, offering a modular *toolkit*-type approach, intended to fit into and provide an evolution path for operators.

7.1.4.3   The office and the factory meet model

Operator service environments have facets of the office – where services are managed, measured, billed and assured – and the factory – where services are produced and where delivery of services is enabled and optimized.

Tying the office and the factory together in an effective manner inside the operator business is vital to operator efficiency, speed and agility. It has also proven to be quite difficult.

There are a number of reasons why this has presented and continues to present difficulties; firstly there is a traditional and somewhat natural impedance mismatch between the technologies deployed in the factory and in the office. This mismatch raises integration barriers between the office and factory. Secondly, the rapid evolution of technology at factory level has made the factory environment a moving target from the perspective of integration with the office. In recent years, a few developments have started to contribute to an improvement in prospects for collaboration between office and factory.

The appreciation of the importance of horizontally structured architecture in the operator environment has led vendors of systems destined for deployment on the operator factory floor as well as the designers of operator solutions to design and build-in functions and interfaces that ease integration with the business process supporting systems in the operator's office environment.

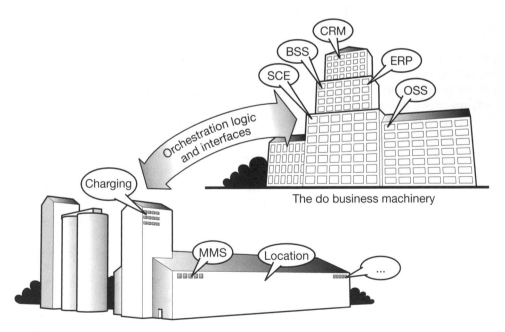

Figure 7.14 The operator service environment – part office, part factory.

Another positive contributor to this area is the notion of the service-oriented architecture (SOA) and the associated practical technologies, such as web services, that enable it. A central tenet of SOA is that systems cooperate over interoperable, discoverable interfaces and that such interfaces are primary objects in the architecture; systems are less tightly coupled with each other as long as they fulfill the contracts of the interfaces. The spread of SOA concepts in real systems should ease the integration pain between office and factory.

A relatively new, but quickly maturing, branch of software technologies which we will loosely refer to as workflow technologies here, already plays a steadily increasing role in forming the glue between office and factory systems. Broadly speaking, these technologies provide development and run time environments with built-in support for interfaces and systems commonplace in enterprise (office) architectures, therefore supporting the creation of the business logic *fabric* between sets of systems, business logic that can *orchestrate* the behavior of office and factory in whatever way the operator's business needs dictate it.

These technologies often offer at least some degree of *codeless* development that, when correctly combined with systems offering interfaces on the right level of granularity, allow the operator's service environment to become more *agile* in terms of the function of its business logic – promoters of this type of architecture refer to this as the ability to *rewire your business every morning!*

7.1.4.4 The internet model

The Internet is the premier example of a successful electronic service environment, and no treatment of a mobile service environment model could be credibly attempted without having a brief look at the Internet model.

The first, perhaps superfluous, statement is that the Internet itself is already part of the mobile service environment. This is obviously true from the technological perspective where Internet technologies either form a base or at the very least inspiration for mobile service environment technologies, but it is also true from the business perspective where Internet-based content and actors form an ever growing part of the mobile service environment landmass.

On the Internet, a considerable amount of intelligence is expected to be available at the end-points (such as devices), allowing the network to focus on the business of optimally routing and transporting packets. In this way, the internet architecture strives for a balance between network-side and device-side function and responsibility that is somehow *just right* at any given point in time. In the mobile service environment, the relationship between device and network side functionality has traditionally been in tension as the industry strives to match the need for flexibility with the market expectations on usability and reliability. It seems likely that the future of the mobile service environment lies with a devolution of processing power and functionality towards devices and a clearer delineation between device and network. It remains to be seen how this can best be achieved while fulfilling the expectations of the user, expectations that demonstrably differ from those of the Internet user.

Although there are similarities between the mobile service environment and the Internet, there are also important differences and it is important to keep these in mind lest we jump all too quickly to the convenient conclusion that the mobile service environment is really just about mobilizing the Internet. Such a conclusion, however compelling or appealing, trivializes and betrays the true complexity of achieving a successful mobile service environment.

One important difference is that mobile clients have historically had less storage and processing capability than their desktop cousins. Increased intelligence in the network has traditionally compensated for this and played a larger role in the overall architecture of the mobile service environment. Additionally, the device population is widely more diverse, with an ever-changing array of form factors, capabilities and what appears to be a constant defiance of the establishment of the lowest common denominators that have been an enabling factor in the fixed Internet world.

Another vital difference is that the nature of the wireless environment is different from that of the fixed scenario for which key Internet protocols were architected and optimized for. TCP in fixed network environments interprets lost packets as a sign of congestion and as a counteraction reduces rate of transmission. In wireless environments, packet loss is a more *natural* fact of life and is often due to events like a user walking or speeding into a tunnel. Reducing the rate of transmission in such situations is a waste of resources. While applications built for the fixed Internet can take a more or less *access agnostic* stance, their mobile service environment cousins cannot yet afford this luxury: they have to adopt a more *access aware* stance, tolerating shorter or even longer coverage interruptions in a graceful manner.

7.1.5   Service environment standards

Standardization plays a vital part in the end-to-end technical viability of the mobile service environment and a number of standardization bodies are creating and specifying technologies for use in the mobile service environment. As a serious treatment of the efforts of these bodies would warrant at least a full separate chapter, we focus on enumerating and introducing the role of the most important standardization bodies here.

7.1.5.1   3GPP (the third generation partnership program)

3GPP is the main body for standardization of WCDMA, UTRAN and the evolution of GPRS and GSM including aspects vital to future developments such as IMS. 3GPP is also responsible for standards related to partner (content/service provider) access such as open service access (OSA) and to user data such as generic user profile (GUP). 3GPP2 is the main standardization body for CDMA2000 and CDMA evolution.

7.1.5.2   OMA (Open Mobile Alliance)

OMA is a main body for service layer standardization in the mobile domain, with participants from the telecommunications and IT industries. OMA has proven itself successful in recent years in standardization of more bearer-agnostic service standards requiring telecommunications and IT participation acceptance to succeed. OMA is responsible for standardization in multimedia messaging services, location services, digital rights management, device management, instant messaging, presence, push-to-talk over cellular, and many more. OMA places significant focus on ensuring service interoperability across all parts of the service environment.

7.1.5.3 IETF

IETF (the Internet Engineering Task Force) is an organization that works with the evolution of the Internet architecture and the smooth operation of the Internet. The IETF plays a vitally important role in the mobile wireless industry as IETF standards are introduced more widely into mobile networks and mobile networks strive to offer Internet-like services to mobile subscribers. 3GPP IMS, for example, is to a large extent based on work in IETF.

7.1.5.4 The parlay group

Parlay is an open multivendor consortium formed to develop open application programming interfaces (APIs). The work in Parlay is driven in close cooperation with 3GPP (that has a tightly related area called OSA) and ETSI TISPAN. Parlay standards and architecture are an important part of opening up the network capabilities to the community of content- and service-providing entrepreneurs viewed as crucial to the emergence of an energetic mobile service environment marketplace.

7.1.5.5 W3C

The stated goal of W3C (the World Wide Web consortium) is to develop interoperable technologies to lead the web to its full potential. W3C is behind the development of many of the standards considered to form the *lingua franca* of the Internet; examples are HTML, DOM, CSS, HTTP, PNG, SOAP and HTML. In an environment so clearly inspired by Internet and Internet-like technologies, the work of the W3C is incredibly important to the end-to-end function and quality of the mobile service environment.

7.1.5.6 JCP

JCP (the Java Community Process) is the main organization for standardization of Java and Java-related interfaces. Java is clearly of importance in the network-side service environment systems such as application servers. In recent years, however, Java has also become an important enabler for customizing and extending device capabilities via the J2ME technologies that form the base for richer device-side applications such as games, connected-productivity clients (calendars, etc.), news tickers and so on, and have achieved wide support in many devices in recent years.

7.1.5.7 WS-I web services interoperability

WS-I is one of the leading web services *de-facto* standards organizations. As web service technologies continue their ascent towards being the top choice for enabling the *opening out* of the service environment, WS-I standards have the potential to take on strong meaning for the mobile service environment.

7.1.5.8 OASIS

OASIS (Organization for Advancement of Structured Information Standards) is a consortium that drives development, convergence and adoption of e-business standards; it produces web services standards along with standards for security, e-business, and standardization efforts in the public sector and for application-specific markets.

7.1.5.9 LAP

LAP (the Liberty Alliance Project) is an organization that creates standards for Federated SSO (single sign-on) and permissions-based user attribute sharing within a federated community (telecommunications, manufacturing, e-government, e-health). LAP aims to integrate its efforts with other mainstream web service standards emerging from other standard bodies. As the mobile service environment opens out into its own flavor of a federated environment, technologies like those from Liberty are likely to be in greater demand.

7.1.5.10 TMF

TMF (the TeleManagement Forum) is an organization dedicated to operations systems support (OSS) communication management issues with the goal of improving the management and operation of information and communications services. TMF is responsible for standardization such as eTOM (enhanced telecom operations map), which is a widely used and accepted standard for business processes in the telecommunications industry. The eTOM describes the full scope of business processes required by a service provider and defines the key elements and how they interact.

7.2 A SERVICE ENVIRONMENT WISH LIST

In the sections above, we have taken a whirlwind tour of the mobile service environment; we looked first at the environment as a marketplace, then later as a machine serving that marketplace. We devote a substantial amount of the remainder of this chapter to aspects of service environment design, but before we dive into that area, we pause and take stock of what we believe are some key service environment qualities given what we have discussed so far.

A typical operator request for information, proposal or tender in the mobile service environment today will run to several hundreds of pages and include thousands of functional and nonfunctional requirements. Technical responses to those requests can be summarized in terms of kilograms of specifications and blueprints, rather than pages! In this light it may seem rather futile to attempt a summary of mobile service environment needs in the limited space available here; however we believe that there are things that are so important that they merit highlighting, things that may often disappear amidst the noise of all those detailed requirements and specifications, things that weigh only a few grams out of the many kilograms in the average request for information/request for proposal (RFI/RFP) response.

7.2.1 Actor related

One of our opening statements on the service environment was that it was *about* people; the question we always need to ask ourselves is what those very people might expect of the service environment.

7.1.2.1   Users

The user wants advanced, cool or fun features accessible in a simple manner, reliably available at a low cost and high speed. Since the user is the most important service environment user, then it probably goes without saying that one quality pervading all service environments should be user focus. The problem, however, is that it usually does go without saying and end-to-end service environments are generally not designed with this simple mantra at their center. Certain parts of the service environment, such as the devices are highly user focused; others are less so. What is needed is that all parts of the service environment place the user center stage.

Let us look at some of the words used here: accessibility starts with the device, and today devices often need configuration in order to enable access to services. Configuration work that adds value for the user, such as that used to personalize the device to the user's own distinct tastes via ringtones, covers and so on should be seen as *good*; configuration that is of a purely technical nature should be seen as *evil*. Device and enabler designers should strive for zero configuration and, where this is impossible, the service environment should offer automatic or semiautomatic device configuration. In the interim our service environment actors may be best advised to adopt a stance where devices are fully configured and open on purchase and where the user can selectively close services if desired, rather than today's stance where devices are often sold with a stance of closed services that are selectively opened.

Simplicity pervades the environment; As we saw in Chapter 5, if 9 out of 10 things are simple but the tenth is difficult, then the experience is difficult in overall terms. We looked at how devices can contribute to simplicity; the operator service environment can contribute by supporting features such as single sign-on, partners can contribute by accepting operator assurances of identity, and so on.

Reliability also pervades the environment end-to-end. Devices must be reliable, applications installed on devices must be mobility aware, networks and operator mobile service environments must be engineered for reliability and availability, and operational procedures must contribute to the assurance of service availability, and quality of delivery. An otherwise appealing, accessible, and simple service environment will fail to attract if it is perceived to be unreliable. While other factors can and should be designed in, reliability requires constant hard work and discipline in areas such as service assurance. Perhaps we should have added the statement 'The service environment is about hard work' in our introduction!

Low cost of service production is a vital quality required in service environments. If the cost of creation, production and service delivery is high, then entrepreneurs will not be able to afford to even create services, users will be unable or unwilling to purchase them and operators and others will not be able to make a business out of offering them. It is, unfortunately, quite possible to create a service environment that is so rich in capabilities and functions that the cost of service production is higher than the price the market will bear.

Low cost of production is a survival factor. Meeting it is difficult but essential; each part of the service environment must do *just enough* and simplicity must be kept at the heart of the designs, especially in the network parts of the environment. Low cost of production must be a design factor in products, services and operations throughout the service environment.

High speed: just like horsepower and torque in the car market, or gigahertz values in the computer market, our work in service environments is often obsessed with bandwidth and less concerned (on the surface) with latency, as we have seen in earlier chapters. The best way to get lower latency in the service environment is to put fewer things in the way of the traffic; it can of course be argued that certain things such as bearer level charging or service authorization must be in the traffic path and in many cases this is true. The important thing, however, is that we organize our service environment in such a fashion that these things occur the minimum possible number of times. As a multivendor, multinetwork, multiactor industry we must avoid the prospect of a multilatency situation, where each leg of the traffic path, each product or even each departmental organization unduly adds to latency, directly translating to user dissatisfaction or user loss. We can improve latency by applying some technologies; however the organizational component remains dominant – putting something in the way of traffic slows it down, and putting more in the way of traffic slows it even more.

7.2.2   Service related

In order to deliver advanced, cool and fun features, the service environment needs to be open and flexible enough to allow a diversity of services to be offered and to thrive. The most important quality in this respect is to avoid placing too many restrictions or obstacles in front of services. Service creation should require only well-known, widely used tools and should demand as little expert knowledge as is rational. Service deployment including deployment on devices should be simple and the service environment should place as few restrictions as possible in the way of such deployments; only specialized services should require deployment inside the operator's environment. The set of end-to-end functions available in the service environment should be relatively stable and sufficient to allow most classes of services to function well. A service that chooses to leverage an exotic function available in one network is likely to be incapable of functioning end-to-end and will have a limited addressable market. Another important quality of the service environment in this respect is that it should be inexpensive to launch and test services. In essence it should be inexpensive to try and to fail; environments where failure is expensive hurt creativity and innovation. In summary, it would seem evident that the mobile service environment has to focus heavily on achieving excellent performance in a few critical end-to-end functions that have universal applicability as well as promoting openness and stability.

7.2.3   Business related

Marketplaces are usually characterized in terms of things like buyers, sellers, goods and prices. What implications does facilitating a marketplace put on the service environment machinery then? We have already spoken about some important aspects above; firstly, a marketplace has to have attractive goods that are to be traded, or at the very least goods that are in demand together with buyers and sellers for those goods. Secondly, a marketplace must provide space or mechanisms where buyers (such as user portals) and sellers (such as operator partner self service portals) can meet and trade in these goods. Thirdly, the marketplace must provide some important qualities such as efficiency (a low relative cost of service production), trustworthiness (reliability, security) as well as mechanisms for credit worthiness (real-time charging), payment (settlement facilities) and so on. Finally, price stability is a quality present in many markets; it helps ensure a sustained engagement of all actors in the market-making process. If prices fluctuate too much, the market loses predictability and actors find it less attractive to participate. One factor influencing price stability in the mobile service environment may be the constant and relatively rapid evolution and injection of new technologies into the environment; new technologies cost money in most cases. In general, the introduction of new technology into the service environment needs to be carefully matched to the marketplace conditions and often technology costs cannot be recouped from the market via higher prices but need to be recovered via increased efficiencies, greater business volume and so on as the marketplace expects constantly declining prices over time. Perhaps our service environment machinery has to focus much more on qualities rather than features in going forward?

There are countless other things we could discuss in our service environment wish list topic; however, as space is limited we will now move on to our next topic, which deals with service environment design.

7.3 SERVICE ENVIRONMENT DESIGN

One way of looking at design is as an act of solving a given problem or set of problems, given business, technological and timing constraints, while at the same time optimizing for a given set of long- and short-term criteria. The art of good service environment design then, is a marriage of problem-solving skills, knowledge and experience, with a careful identification of the real problems and goals that the resulting service environment design should solve.

There is no universal cookbook for mobile service environment design, although there are many valuable sources of recipes; such design is a multidisciplinary effort involving human factors, engineering and technology working in concert against the backdrop of the marketplace needs. Most professional practitioners of mobile service environment design, whether they come from the operator, system integration, service provider, device manufacturer or other background, are likely, out of sheer necessity, to put in place their own personal book of recipes for successful service environment design. The evidence of such recipe books is visible in the shape of solutions, products and in publicly available blueprints for such theoretical or real environments. In this section, we turn our attention to a selection of design topics, a few of the recipes that might appear amongst the large number of dishes in any mobile service environment design cookbook.

7.3.1 Presentation and rendering

The notion of a portal acting as one of the storefronts on the main street of the service environment is one that is well known from the public Internet and one that made an early

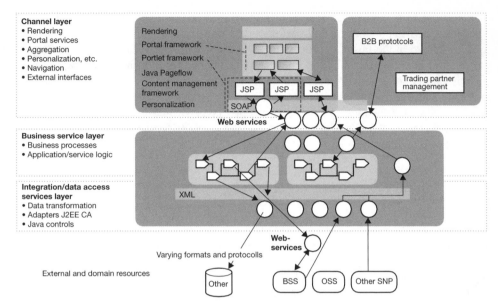

Figure 7.15 Logical composition of a portal architecture.

migration into the mobile space. In this chapter, we take a quick look at the anatomy of a modern mobile portal architecture

The portal architecture shown above is divided into a number of layers: the channel layer is used for the interaction towards a portal user, and the business services layer controls business processes and application logic. The integration/data access services layer handles data transformation and communication towards external entities (other than portal users). We will take a look at each of these in turn below.

Channel layer: this layer controls entry points to the portal via different channels. For users, the channel is normally HTTP-accessed mark-up language-based interaction (i.e. browsing). For other actors such as business-to-business interactions, a web service-based channel may be offered. A third channel might be an interface to SMS-based services offering an entry point for incoming SMS-based requests, such as for service activation and so on. The channel layer is responsible for the user and device adaptation of the portal content. It defines all user interaction with the portal such as the navigation logic. It also defines the look and feel supporting things such as themes and skins.

Business services layer: this layer supports the portal business processes and workflow and is separated from the strict presentation and navigation support offered by the channel layer. This architecture is similar to an application of the model view controller (MVC) pattern to a portal context. Presentation is clearly separated from business processes and those business processes can and should be shared by all presentation (i.e. all channels). In recent years workflow technologies have become popular in this area, not only because they contribute to a more rapid development style but also because they explicitly enable business processes to be created that are a mix of automated and manual procedures and enable the coordination or orchestration of the systems involved in realizing the business process.

Integration/data access services layer: this layer facilitates all communication with adjacent systems and enablers and performs data transformation to and from formats and models defined in the business process tier and elsewhere. One example of such usage is when a user is registered and basic information is fetched from business data repositories, such as enterprise or network databases. This layer often uses a connector style approach to enable and integration layer that exposes a set of stable service or interfaces to upper layers, such as the business process and channel layers, while using pluggable modules called connectors to implement interfaces to the *backend* systems, which this layer is intended to integrate towards.

7.3.1.1 Handling device diversity in mobile portals

Device diversity is one of the main difficulties encountered in creating working and compelling mobile portals. Luckily, some technologies have been developed that support us in solving this problem by rendering a single source of content for the many devices in existence. Delivering content to multiple devices means overcoming differences in device capabilities, including the supported device: mark-up language support (HTML, WML, XHTML MP, CHTML), media capabilities (images, flash, audio), memory restrictions (e.g. the size of content a device can receive) screen size (available display area) and so on and so forth.

The range of devices that must be supported can be simplified somewhat by organizing broadly similar devices into device classes, for example, PC browsers, PDA and mobile phones. Creating content for these three classes is more manageable.

However, the *mobile phones* device class will encompass devices that require a range of mark-up languages, such as WML 1.x, WML 2.x, XHTML MP, CHTML and HTML 3.2, so extra device classes have to be created to cater for those different mark-up languages. In addition, content has to be created for each of the device classes we have identified so far. To complicate the device classification problem further, each device within these classes will in turn have differences in attributes such as display capabilities. These differences affect the kind of content we can deliver, such as which media type to deliver, (GIF, JPEG, etc.) and how to tailor content for the device's screen/memory size.

There are several approaches to solving the problems associated with delivering content to multiple devices. Each solution must address the different device classes, as discussed above. A first, obvious approach is to create and maintain multiple sources of content, one for each of the different device classes we have defined. Another method is to use XSL stylesheets that require us to define the initial content in an XML format and use a separate stylesheet to describe the transformation for each of the different device classes. Both of these methods require multiple files, one for each device class. Then, as individual devices have slightly different requirements and capabilities, we are required to modify our files to cater for these differences. This requires the addition of more intelligence in the method of transformation and/or adding extra sources of content, which in turn creates a significant maintenance task.

A second approach is to add more comprehensive technological support for the problem. We can start by adding a data store containing all the device capability information, including the device's unique content requirements. We can then add the intelligence to identify the user's device (or nearest matching class) and to tailor or render the content to best suit that device. If we are to add such support then it is essential that we leverage the existing design knowledge of the web design community as well as integrate with the most common design and run time tools, for this reason it is best to use standardized mark-up languages such as HTML/XHTML (the W3C standard for defining web content) as the basis for defining the presentation of content. The rendering engine can then automatically transform the HTML/XHTML for each of the different device classes into the required mark-up language, such as XHTML MP, HTML 3.2, CHTML and WML. Final tailoring of the automatically rendered content can also be added in order to address the slight modifications required by different devices in a device class (e.g. images capability and table support).

Later, we will take a look at ways in which the device configuration burden often placed on the user can be alleviated using a concept termed 'automatic device configuration'.

7.3.2 Content delivery support functions

We can use content lifecycle as a means to further illustrate the functions of the content delivery support functions, which we introduced a little earlier.

Once content (such as a ringtone or movie) is created, it is injected, submitted or ingested into a content management system where it is generally subjected to some form of editorial control and approval.

After the approval, which is often supported by a workflow type system, that same content may be encoded into one or more alternative formats, for example, a music file in MP3 format could be encoded/transcoded in a number of alternative formats to enable it to be downloaded and used in various devices, or a video file might be encoded in various bit rates, resolutions or a combination of these to support optimal delivery over one or

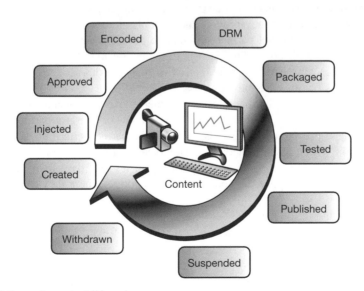

Figure 7.16 Example content lifecycle.

another network channel or to one or another class of device. In addition, appropriate DRM mechanisms can be applied to the content in order to protect the rights of the copyright owners.

Content items may also be linked into one or several service offerings, to one or many customer segments, tagged with different pricing information, marked as suitable for delivery or even discovery on different device types. This linking can be called packaging and is also supported in this part of the service environment. Management of content storage, indexing and searching as well as rapid content retrieval is also supported in this part of the service environment.

Before the content can be published and accessed by users, it is generally tested and its correct function for the targeted channels and devices is assessed using some staging platform, test users and so on. Finally, content is published, i.e. it is made accessible to users via the intended channels. Once published, it is necessary to enable suspension of content, for example in cases where content proves faulty or inappropriate. Finally, all content has a useful shelf-life, after which it should be withdrawn.

In addition to the actual management of the content according to the needs implied by the content lifecycle, this area of the service environment may also contain content delivery-enabling or -enhancing functions such as proxies or servers that optimize partic-ular content types via, for example, compression or tag-reduction, or systems that speed content delivery by, for example, facilitating caching of content closer to edge access and delivery points and keeping such caches synchronized.

7.3.3   Partner support functions

We introduced the partner support functions earlier, just like the content management functions we can use the concept of a lifecycle to explore some the functions present in this area. The story starts when a service provider or content provider registers as a partner of an operator. As a partner they provide their details as required by the partner support functions, often at a dedicated partner self-service portal, and receive acknowledgement of their registration/application. The partner's application may first need to be approved together with other, sometimes manual, activities such as contractual/legal, as part of an operator workflow or other business process.

Once approved, various other operator environment systems may need to be provisioned with information such as the partner's security credentials or operator/partner service level agreements. These procedures, in essence, activate the partner within the service delivery equipment chain, which includes network enablers and partner connection gateways (such as parlay gateways). Once activated, the partner can register their own services. These services, once approved, appear on channels such as portals where they can be subscribed to by users.

Users subscribing to the partner's services may have their devices configured automat-ically if special configuration is needed to enable end-to-end delivery of that service. In addition, the partner's service environment may be notified (via appropriate B2B interfaces based on, for example, web service technologies) that a particular user has subscribed to one of their services, thereby enabling the partner to do any necessary activation/provisioning for the user in their own service environment machinery. Such

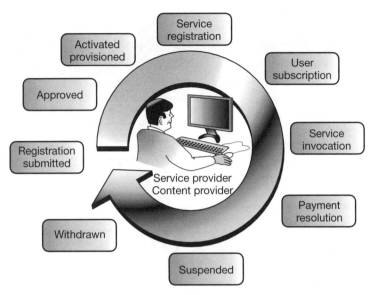

Figure 7.17 Example partner lifecycle.

notifications may transmit user aliases to the service provider if the user's true identity (e.g. their MSISDN) is to be protected.

An activated, provisioned partner with registered services may invoke services using technical interfaces such as the MM7 interface that can be used to send MMS to users, Parlay call control interfaces that can be used to initiate and manage voice calls, or even using operator, regional or special interest community-specific interfaces specially designed in order to support this area. It is important to note that traffic not only flows from the partner to the operator's network but also in the other direction from the user's to the partner's service environment and that, therefore, service invocation is bidirectional and that facets of the user such as their sign-on state, device capabilities and other useful static and dynamic attributes often need to be transferred across service invocation interfaces to create truly useful and usable services.

7.3.3.1 Partner connection gateways

Modern operator service environments often include specific support at the *edge* of their environments, allowing content and service provider access to network enablers and other services. Such supporting functions are often called network resource gateways or service access gateways. We use the term 'partner connection gateway' in this chapter.

The essence of the partner connection gateway is that it offers interfaces to content and service providing partners. In Figure 7.18 we have divided these interfaces into a few different categories based on what is seen in real deployments.

Standardized interfaces are those that are standardized in efforts, like Parlay and OSA or in Parlay-X. These interfaces support services such as call control and messaging and are available in CORBA or web-service flavors. Either technology is addressable from a wide variety of development and run time environments, although web services are more

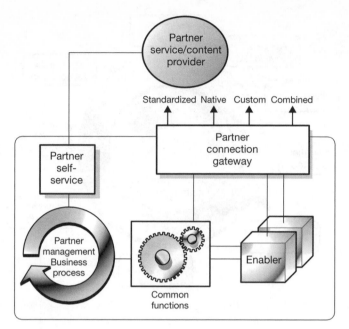

Figure 7.18 Partner connection gateway and surroundings.

popular these days and have a shallower learning curve and more ubiquitous support in development environments.

Native interfaces are interfaces that are offered directly as part of enabler specifications such as the VASP interfaces in MMS (MM7) or in location-based services (MLP). These interfaces are also standardized interfaces; however they are separated here as, when taken as a group, they do not form any unified interface framework whereas OSA/Parlay and Parlay-X do.

Custom interfaces are interfaces that are developed especially by the operator, a partner or a community of operators and partners. Custom interfaces can occur on a regional level or as part of a vertical segment (such as transportation logistics).

Combined interfaces are higher-order interfaces that combine the functions of a number of other standardized, native or customized interfaces to implement newer, richer services. Such interfaces generally require some orchestration facilities in the gateway itself or in service creation environments; they also require that the underlying interfaces can be orchestrated.

The partner connection gateway sits on top of the existing operator service environment enablers and/or functions in the circuit- or packet-switched core or IMS network, which are used to assist in delivering the functions expressed in the gateway interfaces. Additionally, the partner gateway is integrated with the operator service environment business processes for partner management, which are used to drive common provisioning and activation of partner information in the gateway, provision partner and/or service-specific SLA to the gateway.

The partner connection gateway may have one or more other roles. Firstly, many gateways will offer some type of *network protection*. First and foremost is the policing of access to the gateway functions on the interfaces; only duly authenticated and authorized service providers are granted access to the interfaces. Secondly even authorized service providers cannot be allowed to operate the interfaces in ways that might damage the network or impinge upon others' access to network resources; therefore, gateways often support some form of interface invocation or throttling which prevents network overloads or service providers exceeding the terms of their agreements. Secondly, as all traffic on this layer is visible to the gateway, it is in the best position to generate information for both the operator and the partners on partner and user utilization of services for business intelligence or settlement purposes.

7.3.4 Common support functions

As discussed earlier, there is a benefit to breaking out commonly used functionality, most often supporting key operator business processes into separate systems and making them available on open stable interfaces mandated for use by all other concerned systems. The most common examples of such functions are:

- common user data/profile;
- common provisioning and activation;
- common charging;
- common operations and maintenance;
- common identity management;
- common authentication, authorization and session management.

The common functions and the associated protocols and formats via which they offer their services provide a framework onto which all other systems should attach. In essence, then, the common function architecture is best viewed from the perspective of the systems themselves. In Figure 7.19 we look at an example of how *any system* is expected to adhere to a number of interface specifications provided by a common function *framework*. If *any system* is to be a good citizen in the common function framework, it is expected to behave as follows:

- *Any system* should use the diameter credit control application protocol for all real-time charging and should leave things like rating, discounting and such to the common function framework.
- *Any system* should use Radius accounting or FTP together with a mandated CDR format for offline charging.

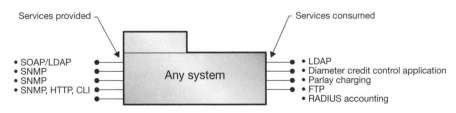

Figure 7.19 Any system and common function interfaces.

- *Any system* should access all user-, service- and subscription-related data that it needs using the LDAP protocol from the common function framework's common directory server and according to the data model/data dictionary defined in the common function framework.
- If *any system* stores its own data related to any of the entities defined in the common function data dictionary, it should implement a provisioning contract that allows the provisioning engine of the common function framework to master (create/read/update/delete) that data (using an SOAP-based interface in this example)
- *Any system* should integrate its fault management and performance measurement initiation and collection via the SNMP interfaces defined by the Operations and Maintenance (O&M) parts of the common function framework.

Additional common functions are likely to exist in most operator service environments, depending on how far the particular operator has pushed the agenda of horizontal architecture within their service environment. We return to the topic of common functions later in the chapter when we look at the task of refactoring in order to achieve horizontalization.

7.3.5   Service enablers

We touched on the topic of service enablers earlier when we first spoke of the service environment as a machine; there we referred to enablers as pieces of machinery, providing functionality and interfaces towards *lower* network layers in the operator service environment. Each and every operator service environment contains a set of enablers that are the engines of vital services such as MMS, WAP, mobile positioning and so on. These same enablers, however, also extend beyond the walls of the operator service environment to the devices and to the service environments of other actors such as content providers. For example, the MMS enabler has an implementation in the operator network (MMS-C) and on the user's device (MMS client), and can even extend to the content provider service environment (via MM7). Taking the above into account, we can look into the world of enablers from two different angles, enablers as *machines* and enablers as *specifications*.

7.3.5.1   Enablers as specifications

The notion of enablers as specifications, as opposed to the more usual notion of enablers as machines, has become prevalent in recent years. OMA, for instance, refers to enablers as 'interoperable components that enable the interaction between different components and applications developed by different providers (e.g. device and network suppliers, information technology companies and content and service Providers).' We present a selection of enablers from OMA as a base for illustrating the notion of enablers as collections of specifications that literally *enable* the realization of interoperable end-to-end functions in the mobile environment.

The enabler specifications presented in Table 7.1 affect devices, content providers, operator environment machinery, end-to-end traffic/signaling or a combination of all these.

Table 7.1 A brief survey of OMA enablers

OMA enabler	Description of the enabler
OMA Browsing	This enabler specifies browsing capability (including WAP) for mobile devices and associated network services (e.g. gateways and proxies). The browsing specifications draw inspiration from internet technologies, adapting them, where beneficial, to the mobile context.
OMA Client Provisioning	This enabler specifies mechanisms, such as over the air (OTA) provisioning, for remote configuration of WAP clients. This eases the configuration burden on users as well as enabling operators and service providers to improve their customer service.
OMA Data Synchronization	This enabler specifies mechanisms for data synchronization between devices and other systems using technologies such as SyncML
OMA Device Management	This enabler allows an external party, such as an operator system, to remotely set parameters, conduct troubleshooting/servicing and install or upgrade software on mobile devices, again SyncML based technologies are used.
OMA Digital Rights Management	This enabler allows content copyright owners to protect their rights by defining mechanisms for attaching cryptographic protection to media objects (games, ring tones, photos, music clips etc.) as well as specifying the conditions for how the content can be unlocked and used by the receiver, shared with others and so on.
OMA Download	This enabler specification deals with mechanisms, such as application level protocols, necessary to enable confirmed download of content by users. The ability to confirm that premium content has been successfully delivered to the user is key to the billing for such content.
OMA Email Notification	This enabler defines mechanisms allowing e-mail servers to notify mobile device e-mail clients, of e-mail message events, using WAP push technologies.
OMA Games Service	This enabler discusses functionality for session management, connectivity, metering, competition management and logging with a view to easing portability and improving interoperability in the mobile games space.
OMA Instant Messaging & Presence	This enabler provides for the definition of specifications for instant messaging and presence services in the mobile domain as well as to and from Internet-based instant messaging services.

(continued overleaf)

Table 7.1 (*continued*)

OMA enabler	Description of the enabler
OMA Mobile Location Protocol	This enabler specifies a protocol for obtaining the geographical position of a mobile device in a manner that is independent of the underlying network technology.
OMA Multimedia Messaging Service	This enabler specifies the well-known MMS service including device and network side elements and protocols, VASP interfaces; inter operator roaming and so on.
OMA Online Certificate Status Protocol	This enabler specifies mechanisms for validating that a certificate and it's associated private key are still considered trusted and valid.
OMA Presence Simple	This enabler specifies mechanisms for managing the collection and controlled distribution of presence information in mobile network environments. The specifications provide a client server framework, presence sources and watchers, as well as presence and list servers.
OMA Push To Talk Over Cellular	This enabler specifies the POC service, which is a walkie-talkie inspired voice communications service, allowing users to communicate with individuals or within groups.
OMA Standard Transcoding Interface	This enabler specifies an interface between a transcoding platform and application platforms. Applications can request transcoding of media files based on specified transcoding parameters, device capabilities and so on. The transcoding platform does the actual transcoding work, returning the desired object to the application.
OMA User Agent Profile	This enabler specifies mechanisms enabling the flow of a User Agent Profile (UAProf) between the device, intermediate network points, and origin servers. Origin servers, gateways, and proxies can use capability information to ensure that the user receives content tailored for the device on which it will be presented.
OMA Web Services (OWSER)	This enabler specification is really two separate specifications 1) Core specifications: specify basic web service infrastructure for providing/consuming web services in an OMA environment, and 2) Network Identity specifications: define components needed to provide important aspects of the network identity related capabilities in the OMA Web Services specification.
OMA Wireless Public Key Infrastructure	This enabler specifies security mechanisms based on public key technologies. These mechanisms enable secure authentication, ensure confidentiality and protect the integrity of messages exchanged between systems that use the specifications.

Additionally, some of the specifications result in the creation of machines, in the traditional sense of the word 'enabler', that are subsequently deployed in operator service environments.

7.3.5.2 Enablers as machines

Traditionally, the word 'enabler' has been used to denote the concrete machinery for delivering a service in the operator environment. There are a good number of such machines that are not strictly covered by OMA specifications of *official* mobile service environment enablers but are nevertheless vital parts of the operator service environment plumbing today. Examples of such enabler machinery are SMS gateways, game servers, Java download servers, ringtone servers, streaming servers, optimizing proxies, corporate e-mail mobilizers. A good many of these *enablers* implement standardized interfaces and protocols; however, not all of them are *standardized* enablers in the OMA sense. The presence of such enablers is likely to continue well into the future; as OMA itself states, the word 'enabler' can be used to denote any component solving a common need, reusable and reused across environments.

7.3.6 Single sign-on

Studies show that each authentication challenge made to the user leads to a successive *thinning* out of the user population that eventually reaches the service after negotiating each of the login hurdles.

Leveraging network based authentication or any previous explicit user authentication (login) removes the obstacles that prevent users reaching the services and contributes to higher usage and better user satisfaction.

Historically, each application environment has provided its own mechanisms for determining the identity of a user (authentication) and deciding if and when a user is granted

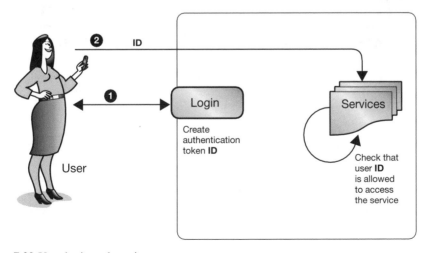

Figure 7.20 User login and service access.

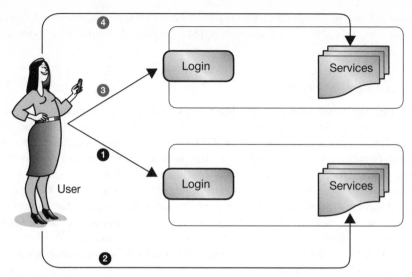

Figure 7.21 Multiple authentication.

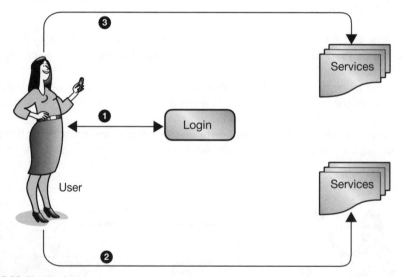

Figure 7.22 Single sign-on.

access to a service (authorization). Usually, these mechanics are hidden behind a *login application*. The user provides their name and password to the login application, which generates an identity token that the user can supply as identity proof to services. However, typically each application environment has implemented this login procedure differently. This means that a user must reauthenticate every time they switch between different application environments.

SSO is a convenience feature that enables users to access different services or service providers (SPs) without authenticating at each one. In other words, a user authenticates

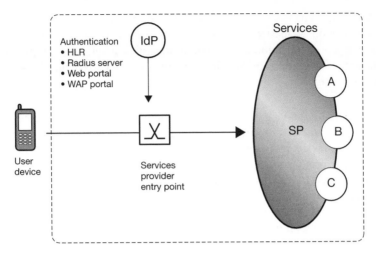

Figure 7.23 Basic single sign-on.

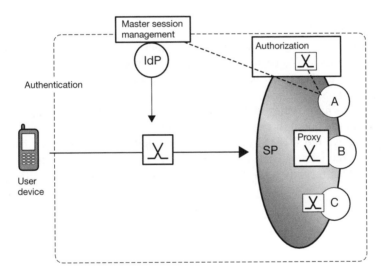

Figure 7.24 Authorization and session management.

only once and the resulting authentication is valid for entrance to all other services or SPs. SSO simplifies access to services for the user. The simplest form of SSO allows users to request services once they are inside the operator service environment (walled-garden), requiring no further logins. Users need to prove their identity with the help of an identity provider (IDP) to pass the entry point that controls access to an SP's network, thereby satisfying one of the requirements to be authorized to use some services, for example, a web login screen blocking access to services behind.

In the example shown above, the operator plays the role of both identity/authentication provider and service provider. Authorization adds the possibility of controlling the right of subscribers to launch certain services. A service provider can now, for example, charge users for subscription. A master session can be used to keep track of the user state in the service layer, e.g. authentication state/strength, identities, event timestamps and timeouts. These data are used to deliver user information to services. Figure 7.24 shows three different kinds of sign-on:

- *Application-based* – service A has a built-in support that it can use to retrieve authorization and user information from the authorization and master sessions, respectively.
- *Proxy-based* – service B assumes that the proxy performs authorization and session management. A proxy service can take over authorization and master session management tasks for services that cannot handle that on their own. The proxy can do this tasks standalone or in cooperation with an external supporting entity.
- *Legacy service* – Service C is not SSO-compliant since it handles authentication, authorization and sessions on its own.

7.3.7 Device configuration

Today's advanced packet and data-enabled devices are designed and developed with little common (i.e. cross vendor) view on harmonized approaches to configuration. As a result, although similarities exist, each device brand and model has its own unique way of configuring settings. The result is a complicated procedure requiring high-level skills for proper configuration.

Many operators have deployed device management over the air systems; users typically enter the operator's portal in order to type in the subscription number and phone model before they receive the OTA SMS. Preconfigured devices are also being used, often to jumpstart new services, but such devices also need additional configurations for new services later. One recently seen development is the attempt to automate the process of

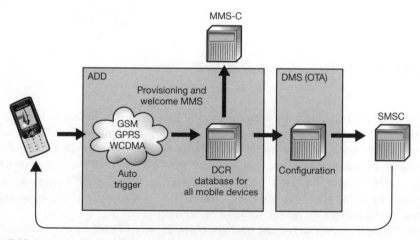

Figure 7.25 Conceptual view of automatic device configuration.

configuration insofar as is possible and to relieve the user of some of the burden inherent in the process today.

The term 'automatic device configuration', or ADC, can be applied to a configuration support system that detects new users and device combinations and automatically configures the device, and optionally the network side service enablers, based on any new states or combinations detected. Such a system can cover the following example scenarios.

- When a user uses a device to attach to the network for the very first time, an OTA SMS with the configuration data, foreseen by the operator, will be sent to that device. The system will keep track of the fact that this configuration has been sent for that user and device combination.
- If the user moves their SIM card to a new device, an OTA message, suitable for that device, will immediately be sent. If accepted by the user, the device will be configured to access all the user's usual services, such as GPRS bearer or MMS.
- If subscription changes are made in the home location register (HLR), for instance if a new APN is added, affected user/device combinations can be immediately updated. This feature can be used, for example, when an enterprise signs a deal giving them Intranet access using a dedicated APN.

An automatic device configuration system can consist of a number of elements; firstly the network is responsible for forwarding detection events or notifications based on the signaling procedures that are part of the networks nervous system – for example, when users attach to the network.

The second part of such a solution is what is called a device configuration registry (DCR). It is capable of comparing the international mobile station equipment identity and software version number (IMEISV) and user parameters received in network notifications with current/previous values in its database and makes a decision on whether the particular device and user combination requires a new configuration to be sent or not. If configuration is required then the DCR sends a notification, including the IMEISV, MSISDN (mobile station international ISDN number) and APN(s), to the DMS, which is the third part of the solution.

The DMS part then configures the device using an OTA SMS after fetching the relevant settings belonging to the APN(s). This OTA SMS may be optionally preceded with a text SMS informing on the purpose of the new configuration. In addition to the configuration of the actual device itself, this automatic device configuration system may also automatically provision network side enablers such as MMS-C.

In addition, such a system can collect data on user device selections and churn over a time based on its detection capabilities. This data can be a valuable source of information enabling tuning, targeting and continuous improvement of the service environment catering to the most popular devices or device usage patterns.

7.3.8 Refactoring for horizontalization

Historically, operator service environments have grown in a rapid fashion under time-to-market pressure by the continuous addition of vertically integrated applications and services.

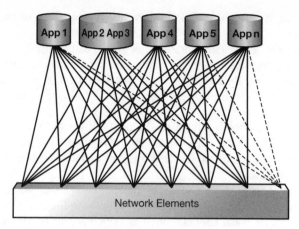

Figure 7.26 A common scene in current service environments.

The main symptoms of this situation are quite severe from the perspective of the operating business and the user experience, mainly:

- service creation and rollout is slow, expensive or both;
- service delivery itself may be unreliable;
- operations are expensive and labor-intensive;
- introduction of new important technologies may be very difficult.

In essence, many service environments exhibit the undesirable qualities of any unstructured environment: unpredictability, inflexibility and an associated high cost of operation that leads either to high cost to the user or lower margins to the operating business. As a result, an important recurring theme in service environment design is that of horizontalization.

In a horizontalized operator service environment, all functions that can be shared among the various services and applications are identified and interfaces towards those common

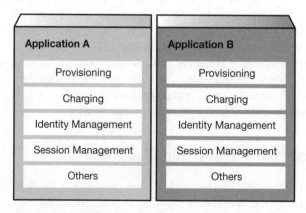

Figure 7.27 Vertically integrated applications.

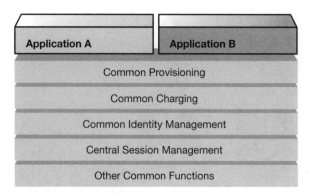

Figure 7.28 Horizontally integrated applications.

functions are defined for the use of each and every application. More often than not, the hottest candidates for common functions are those supporting vital business processes such as charging/billing or provisioning; however, any piece of functionality that is shared across the environment qualifies as a common function.

Rather than begin by presenting a fully horizontally structured operator service environment, we present horizontalization from the perspective of an improvement or reengineering perspective as it is more commonly a concern in such scenarios than in up-front design.

Horizontalization, when done in a refactoring context, is usually an incremental process, proceeding according to the functions that deliver best technical or commercial benefit in a given time frame while creating minimum disturbance to ongoing business operations. Horizontalization is often performed in order to increase the businesses ability to manage its infrastructure, to save costs on operations, to better leverage investment by investing money once on particular shared functions and infrastructure and/or to increase the businesses ability to deal with change in the commercial or technological dimensions. In summary, horizontalization should help to significantly reduce operator OPEX (operational expenditure) and CAPEX (capital expenditure).

Independent of whether horizontalization delivers a benefit to the business, achieving it always incurs a cost. In this sense it is essential that cost–benefit analysis is performed by the business for each and any horizontalization effort planned and undertaken.

7.3.8.1 Data architecture

Data architecture is, in itself, an important part of any general architectural strategy. Coherent data architecture, catering to user, subscribers and service data and supporting the operator's service environment business processes is key to achieving a manageable, performant and flexible service environment. Organized and well-planned data architecture is a key prerequisite to overall horizontalization goals as all common functions are driven by common data, and operator business processes act on that common data.

In this section we will take a look at aspects of solving a common problem seen in many existing service networks: the problem of improving upon the *ad-hoc* data architectures that have arisen from piecemeal integration of various systems over time.

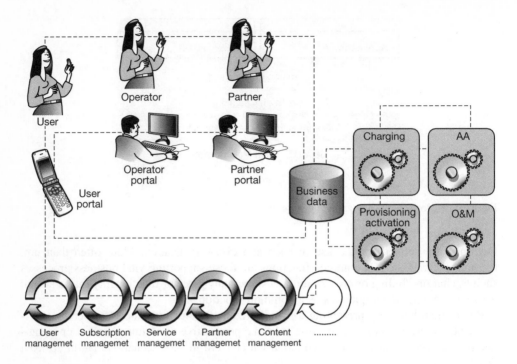

Figure 7.29 Business data links: actor entities, business processes and common functions.

7.3.8.2 A typical starting point

The situation depicted is typical in the sense that it depicts an arrangement of systems that have been created, acquired, purchased and/or integrated into a service network over time. The result is *ad-hoc* data architecture: an aggregation of a set of vendor or operator data models that have some accidental commonalities but often no explicit organization.

New systems are added over time, and are often integrated as is into the network. Furthermore, the systems often come with all that is needed, frequently including data models as well as the infrastructure required to host them. Such solutions may also have the ability to allow/facilitate user-specific personalization and contain persistent user data according to their own local data models. The result of all this growth can be summarized as follows: traffic cases may work just fine, however the data architecture is highly undefined and can be summarized as being;

- An aggregation of all data architectures with no common data keys, access protocols, policy for shared vs distributed data.
- The portal often becomes a *data hub* yielding lock-in to the portal architecture; in addition, the integration of each new system disturbs the existing data architecture, and it is also difficult or impossible to foster ACID (atomicity consistency isolation durability) properties across the data.

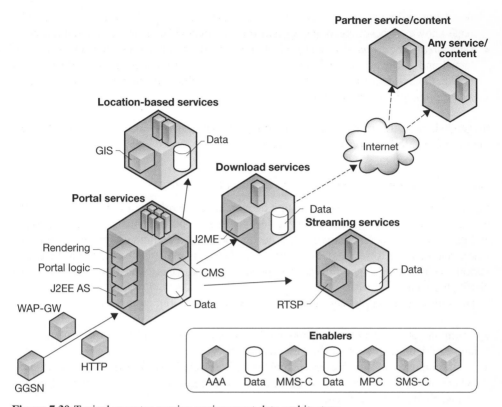

Figure 7.30 Typical operator service environment data architecture.

- Supporting common functions is difficult, as these often require shared definitions/access to shared data:
  - What are the keys in each system and how can they be mapped?
  - Which protocol to access data over?
  - What system stores the data?
  - Where are authentication credentials, policies and certificates?

If we take the example of trying to add support for certain common business processes such as provisioning, we can see that the work avoided during the *organic* service environment growth leads to lots of complexity in the provisioning system. The provisioning system has to know what systems have data, how to contact those systems and what access keys to use to retrieve data. In essence, the system needs to know a lot about a potentially large number of diversified systems in the network and it has to know more and more as the network grows. When this situation arises, some obvious paths forwards for the provisioning system are:

- *Do piecemeal integration* – creating an adapter for each system to be provisioned and creating associated glue logic to handle access-protocol conversion as well as data model translation/mapping. Additionally the 'existence' of each new system must be made known to the provisioning system (or a mediator) by means of configuration.

- *Refactored integration* – an analysis of the typical situation presented above is conducted and a decision is made to lift out certain data to a *common place*, document a data-dictionary/data model for the business and establish a number of preferred data access protocols for the network.

In general, it is the refactored integration path that leads to a solution with the best long-term viability and operating efficiency.

7.3.8.3 Steps toward a horizontalized data architecture

Figure 7.31 depicts a snapshot of the first data architecture refactoring step. In essence the use cases, data needs, business process needs and existing distributed data are analyzed and decisions are made about what data need to be defined commonly for all systems and where that data are to reside. During this step, decisions may also be made on the use of various technologies supporting access, persistence and transactional consistency of the distributed data.

The following example steps can be seen as part of the refactoring step:

- Common data is identified and modeled. The entity relationship model, attributes and keys are defined to support the key business processes.
- A minimum set of suitable access protocols is selected, e.g. LDAP, SQL, XML-WS. Where multiple protocols are needed, an effort is made to implement them all on top of one single data architecture.
- A suitable persistence technology is selected; in general the technology must be transactional capable, support online backup and replication.

The steps of the refactoring process, the resultant choices and actions must be matched against the business costs and benefits. Here we show a possible result of the refactoring

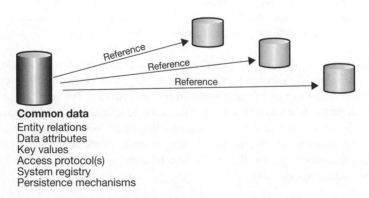

Common data
Entity relations
Data attributes
Key values
Access protocol(s)
System registry
Persistence mechanisms

Figure 7.31 Snapshot of data-architecture refactoring.

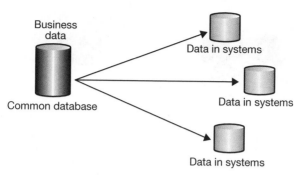

Figure 7.32 Example of refactored data architecture.

process. Note that several approaches are possible depending on the degree to which refactoring can pervade the existing architecture and the degree of willingness to make complete technological choices, both of which are highly situation-specific.

7.3.8.4 Provisioning architecture

Provisioning is primarily about the efficient management of user, subscriber, service and activation-related data and the relationships between those data within the operator service environment. It is intrinsically related to the operator's core business processes and is key to the ability to handle subscriber/service growth and the rapid introduction of new services characteristic of the service environment. Provisioning business processes, if well executed, are a key facet in forming positive subscriber perceptions of operator service and an easy to use network.

In this section we will examine aspects of horizontalizing the provisioning architecture. We take a look at typical *ad-hoc* provisioning solutions that are often seen today, and illustrate how these may be improved upon via a horizontalized provisioning architecture. As coherent data architecture is a significant prerequisite for designing a provisioning architecture, the material in this chapter builds on that previously presented in Section 7.2.8.3.

7.3.8.5 A typical starting point

Typical of today's provisioning architectures are on-site integrations of separate provisioning mechanisms/systems that have been delivered with purchased systems and have evolved over time, often yielding solutions with high cost of operation and well-known weaknesses in data consistency, speed, usability and reliability.

Figure 7.33 shows the customer administration systems (CAS) close to a number of systems and applications that require provisioning and often possess their own local data stores. In the typical *organically* evolved architecture the CAS systems are integrated with the systems to be provisioned via a wide mix of mechanisms, such as GUIs, batch files, command line interfaces and proprietary protocols, and a significant amount of

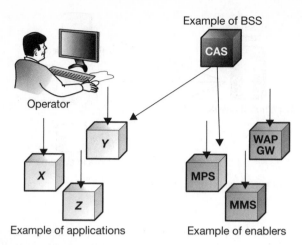

Figure 7.33 Typical provisioning scenario today.

administrator intervention may be required in various provisioning processes. There are some natural consequences of such architecture:

- Since there is no effort to simplify the model, the complexity of the underlying architecture and systems pervades all the way up to the customer administration personnel and ultimately the user.
- Since such architectures often result in a number of manual steps being required in the provisioning process, the cost is increased, the performance is reduced and the probability of errors in increased.
- From the technical perspective there is a significant probability of related data items being out of sync with each other during the potentially elongated provisioning process window. Such inconsistencies may affect actual user traffic and experience of the service and frequently lead to hard-to-diagnose subscriber problem reports, as the inconsistencies are difficult to reproduce.
- As the number of users and services increases, the operator may lose the ability to cope with growth or be faced with the situation where provisioning is too expensive, too unreliable or too slow to cope with the business opportunity in the market.

What is needed is a provisioning architecture that can cost-efficiently and effectively manage the scenario where the potential number of subscribers and services is increasing steadily.

7.3.8.6 Steps toward a horizontalized provisioning architecture

Designing a horizontalized provisioning solution requires analysis and work specific to the operator's unique situation and business goals. However, certain high-level steps are universally applicable:

- Decide upon the key provisioning/activation processes, optimizing for clarity, usability and speed towards the users and in all interactions between administration personnel or systems and users.
- Define/design the data models required in order to support the provisioning business processes. The data model not only supports the business logic in the chosen customer administration system but also provides the framework that reflects service lifecycles and determines how new services and systems are integrated into the provisioning architecture.
- Survey the existing set of systems and technologies deployed in the network with a view to determining further requirements on the provisioning architecture itself or determining and planning the level of change that will need to be made to the existing systems.
- Analyze needs related to known or planned future developments such as the deployment of new network standards, technology changeovers, etc.
- Define/design the architecture required to support the business processes, horizontalized data model and integration of already deployed systems.
- Survey and select appropriate technologies, products and services in order to initiate work towards the desired provisioning architecture.

7.3.8.7   Horizontalized provisioning solution illustrated

This section illustrates a provisioning architecture designed to alleviate the problems arising from typical vertically integrated provisioning solutions. The following are elements of the horizontalized provisioning architecture:

- *CAS* – the customer administration system is business logic and data master for all common subscriber, user and service data. This role relates not only to the service network but also to other networks.

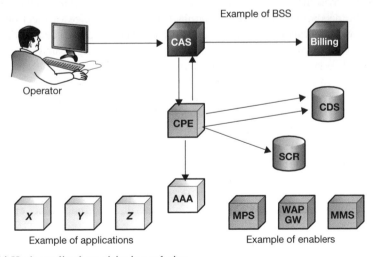

Figure 7.34 Horizontalized provisioning solution.

- *CDS* – the common database system stores user, subscriber and service related data according to a data model. For example, the list of services subscribed to by a user can be found in the CDS.
- *SCR* – the system component registry stores data on the systems that are to be provisioned. Such systems register with it and supply provisioning templates (for example, in the form of XML schemas) indicating attributes and interfaces upon which a system is to be provisioned.
- *CPE* – the common provisioning entity acts as a central point of provisioning in the architecture. It receives provisioning commands from the CAS via an open interface and, by utilizing data stored in the CDS and SCR, can determine changes to be made to CDS data as well as finding out which other systems are to be provisioned and how they are to be provisioned.

The horizontalized provisioning architecture emerges as a framework whereby systems that require provisioning [e.g. MMS, AAA (authentication, authorization, accounting)] adapt to a common data architecture (CDS and data model), register with the provisioning framework (in SCR) and accept provisioning commands via a set of well-defined protocols from common business logic in CPE.

7.4 SUMMING UP

The service environment is about people, and service environment design should be people-centered. We encountered this in a number of areas including accessibility via pre- or auto-configured devices, single sign-on support in the network, keeping network round-trip times for services as low as possible, transparently supporting the myriad devices used by people to access services and creating an environment that is as open and thriving as possible.

The service environment is also about business, and service environment design needs to cater to the needs of a thriving marketplace for the exchange of electronic services amongst a growing number of actors via differing business models. The first directives for the service environment in this respect are the notions of *openness* and *interoperability*; all parties need to be able to technically unite to enable the production and delivery of services. We looked at how partner support functions in the operator environment support the goal of openness; we also saw how one emerging 3G operator is betting their future on what we called the channel provider model. We also found that the notion of *agility*, or the ability to change quickly and inexpensively, is important, and that the notions of *trust* and *fairness* are vital to multiactor business environment viability. Finally we stated that engineering a service environment that allows lowest possible cost of production is essential to the long-term survivability of the business.

The service environment is also, naturally, about technology and machinery. In this respect we presented an anatomy for the operator service environment covering the most important functional areas and discussing their roles and interconnections. We saw that one of the biggest forces acting on the architecture is the notion of *change*. Taking an outside-in perspective, we looked at how the lifecycle of important marketplace goods such as services, content and partner is mirrored by functions in the service environment and how lifecycle is mirrored in common functions, data and business processes in the machinery. We also looked at some specific design islands catering to issues we identified in the marketplace view of the service environment. Finally we looked at the topic of architectural horizontalization in the refactoring sense, how operator service environments can and are being improved via consequent reuse of common functions.

8

Deployment of Services

8.1 INTRODUCTION

There has been much discussion of the design of applications, the architecture of networks and the importance of consumer understanding and roadmap planning. All of this preparation is put to the test when it comes to deploying a service or enhancing a solution with new functionality. This chapter deals with some of the practical aspects of integrating and verifying a solution and preparing it for launch.

This is the challenge – taking a packaged product, often together with several partners or subsuppliers, and actually implementing it. It is here that major headaches can occur, when integration problems will flare up, with various parties inevitably pointing the finger, blaming each other for not supporting 'the standard'. It is also now that the fully functional service can be tried out for the first time. Will the actual end user experience meet expectations? Does the target country's live and loaded mobile network deliver the capacity required? Are there any fatal flaws that were not anticipated in the laboratory? Certainly this is a testing time and this chapter will hopefully provide the reader with insight into how to approach this phase and avoid at least some of the mines in the road to a successful market launch.

8.2 PREPARATION IS KEY

No matter what your role is in bringing a service to market, be it content provider, aggregator, operator or equipment vendor, the true test comes only after signing a contract. The ability to implement a complete solution on time relates very closely to the careful preparation of each participant and precise control and management of the overall project. In this case, the technical merits of the solution are matched by the ability of those responsible to effectively manage a delivery project and to anticipate many of the obstacles that will appear. Many stumbling blocks are unavoidable and impossible to predict, but at the same time, even the simplest precautions can save days and weeks of frustration.

Mobile Media and Applications – From Concept to Cash: Successful Service Creation and Launch Christoffer Andersson, Daniel Freeman, Ian James, Andy Johnston, Staffan Ljung
© 2006 John Wiley & Sons, Ltd

8.2.1   Know the target market

Taking a step back, especially for those players entering a new arena, it is necessary to learn as much as possible about the target market from a business and technical perspective before embarking on an implementation. This can be as simple as addressing a new customer or segment or as complex as moving into new countries.

Following on from an understanding of the market, as was discussed in Chapter 3, one needs to understand more of the practical aspects of operations in a given region:

- The addressable market, especially considering if your business approach requires teaming with a specific local operator. In the United States, 295 million people are split among the major carriers (Verizon Wireless, Cingular Wireless, Sprint PCS and T-Mobile USA) and literally dozens of different MVNOs. This drastically reduces the market that is addressable by dealing with any single operator.
- Rules of operation in specific regions; regions may have strict laws and limitations on business activities – in many regions solution providers are required to possess special licenses for such work and there are restrictions on the granting of such licenses to foreign companies. This can be even worse when deploying in politically sensitive 'embargo' areas. Here, aspects arise such as encryption are issues as well as your choice of third-party vendors included in your solution – even down to the smallest software components.
- Some markets may have limitations when it comes to revenue sharing and subsidizing.

When looking at more of the technical aspects a strong understanding of the target market is also needed when engaged in solution architecture work. Key areas of importance here are:

- Mobile terminals;
- Mobile network capabilities;
- Service environments.

When looking at mobile terminals, the general penetration of devices that support the functionality your service requires (in the target customer segment) is important. Also of importance are the general business practices concerning the customization and sale of mobile devices. In Europe and America it is common for mobile devices to be heavily influenced and customized by local operators, while in Japan, NTTDoCoMo has been dictating terms to device vendors for years in order to strictly control the i-Mode environment. This introduces at least the possibility of launching mobile devices with specific applications and features embedded in the device if you are going to market jointly with an operator. However, in China, operators have no control over the devices released in the domestic market and this means that, as well as not having a definitive idea of the specific devices in use, it is not possible to embed application components on a device other than by using OTA provisioning.

Looking at mobile network capabilities, it is important to be aware of the features of the target mobile networks. Knowing more about the realistic performance and capacity of networks is important, especially when considering bandwidth-intensive services such as video streaming. What network capacity has been dedicated to data services – are there dedicated timeslots in GPRS? Has QoS been implemented in EDGE or WCDMA? Many of these aspects are very hard to find out due to operator secrecy. In the absence of independent surveys on realistic data speeds, the performance of best effort network access should either be manually tested on site in target locations to the guage feasibility of launching a service or within laboratory conditions where you can recreate realistic scenarios.

In addition to pure performance testing, however, it is important to understand the tariff models in place in target markets. For example, one should consider carefully the model for delivering video or music content. Independent service providers operating outside of the operator's walled garden may induce significant volume charges for subscribers that purchase and download large pieces of content. If it is not possible to negotiate flat rate downloads, one must carefully consider the viability of particular services. It has been seen, for example, that MMS delivery of short clips is a viable alternative to straight GPRS downloads in order to avoid the subscriber being charged volume tariffs separately.

On the service environment side it is important to understand the capabilities and features that are available to capitalize on. When considering the launch of a service within an existing portal one must understand the existing service offering and its structure, and consider the impact of new services and how they co-exist and evolve with other existing components. Introducing ringback tones into a portfolio that already includes full-length music downloads will obviously raise discussion about tighter integration and combined offerings – at least from the perspective of the end-user.

When launching services independently one must also carefully consider the feasibility of offering services across the board to all subscribers. Are subscribers blocked from accessing services outside of their operator domain? Is premium SMS the only avenue for charging subscribers? Can you deliver WAP push or MMS messages to all subscribers?

The benefits of due diligence and careful preparation are strongly rewarded when it comes to the more challenging phases of integrating and deploying a new service.

8.2.2 Successful trials

Trials can be a great way to prove the merits of a solution and your capabilities. They can also be extremely costly and time-consuming. Before starting, it is critical to agree on the scope of the trial and specific requirements for success. In addition, one should agree with stakeholders on the course of action to take after the successful completion of the trial. This is valid both for internal trials as well as for external sales and business development.

Strong stakeholder commitment to a trial is important. There are always multiple trials ongoing in a target network and, even though a trial may be funded, it is important that executive interest in a trial feeds down into the rest of the organization since there are always standard operational issues that will draw people away. It is the operational people that will help you to integrate a solution – without their buy-in they will find other day-to-day issues which to them will seem more important.

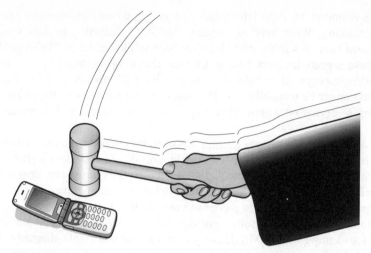

Figure 8.1 Prepare badly for a trial, and you will be judged guilty!

Trials can also suffer and be drawn out in time when delivery teams do not get even the simplest things:

- Easy access to laboratories and trial equipment;
- Dedicated stakeholder contacts and budget;
- Prioritization for planning, integration and demonstrations.

It goes without saying that new products must have been extensively tested before integration, documented clearly and professionally and that all test results are available for those responsible for the target network – without these integration will not occur.

With good preparation and a clear goal for a trial it is possible to get a long way towards commercial launch of a solution. It goes without saying that poor execution and a lack of stakeholder commitment are a recipe for disaster.

It is important to remember that a trial system is just that – a *trial* system. It has normally been designed to provide only the core functions of a service under low traffic conditions and it will never perform as a full commercial system. Using a trial solution as a live product is dangerous to say the least – even if you anticipate low usage or if a target market is still developing.

8.2.3 Understand the business objectives of a solution

It is crucially important to consider the various business intentions behind a new solution and the expectations of all stakeholders such as marketing groups prior to the deployment of a new service.

Consider, for example, how a service will be promoted at the time of launch – there could be a free trial period to encourage people to use the service. In such a case, can the system and underlying network be configured to accommodate such a charging scheme, and will the system handle high capacity demands from day one?

As an example of the capacity limitations in the provisioning solution for an MMS system, imagine that at Christmas time the marketing department decides to run a campaign and makes a special offer to sell 1 million handsets with MMS capabilities on the market. The result is that on Christmas Eve the MMS system will receive thousands of requests to provision new subscribers and the overall performance will go down. New users will (of course) try to send a first picture message to friends and family and the delivery time may increase to minutes and maybe hours, thus destroying end-users' first experiences of the service.

Strong communication and an understanding of the intentions of all groups is important when planning a solution deployment.

8.2.4 Manage expectations – no scope, no hope

When it finally comes down to sitting down and planning the deployment of a solution, it is extremely important to define plans that meet the business objectives of your customer.

It is very important when defining the scope of an implementation that all parties are realistic about what can be achieved. This is particularly relevant in highly competitive environments when there is always a shortage of time, resources and funding.

Given the complexity of solutions, in many cases one should implement only the most fundamental features and even these as a phased approach with each step easily implemented. Define solutions according to the most valuable features that will have impact across the largest possible target segments.

Feature-rich solutions released all at once increase risk and complexity in deployment and promotion and seriously undermine the realistic profitability of a service. Vendors should also be wary of pushing an entire platform in a solution for the sole purpose of realizing a small subset of features.

Consider not only the learning curve of end-users but also of all other links in the chain. If a sales representative is faced all of a sudden with a brand new service with dozens

Figure 8.2 No scope, no hope.

Figure 8.3 Consider carefully the true value of each feature in a solution.

of new options, it is simply not realistic to expect him or her to instantly be capable to promote and explain such features to potential customers. Every phone call back to sales support staff is an added cost that can simply undermine the profitability of a new service from the start.

When it comes to planning the integration, one must absolutely minimize the number of integration points and to try to keep assumptions about terminals, network and service environments to an absolute minimum. When looking at the target network, do not assume that any 'future' features are guaranteed – go with what there is now and limit your dependence on other nodes and systems!

When considering the future, however, scalability is of great importance. While the phrase 'scalable solution' is often abused, it is most important to ensure that a solution that is delivered can handle the load that the business case for the solution anticipated and that this is tested and verified as a part of the integration process – such verification can also highlight dimensioning problems in other areas of the network.

Summarizing the above aspects, the following checklist can be useful for deployment preparation

- ❑ Complete stakeholder buy-in;
- ❑ Define a clear scope with well-defined borders and project timelines;
- ❑ Limit risk by making a phased approach – no big bang;
- ❑ Strip down features to the core functionality that will address the largest possible segment;
- ❑ Minimize the network integration points required;
- ❑ Avoid making assumptions about handset and network capabilities;
- ❑ Do not rely 100 percent in the future features of a roadmap – certain features may change and you must be ready to accommodate this;
- ❑ Make sure your solution is truly scalable and plan to verify it.

8.2.5   Products become solutions

While deployment in the past was very much about installing a product and starting the service, the market now demands more flexible and customized solutions, adapted to specific customer and network needs. This means that these solutions from an integration perspective are very different for each customer. When planning these kinds of flexible solutions you must define at the planning phase the scope for later management of the solution (known as lifecycle management or solution management).

Table 8.1 A comparison of products vs. services and solutions

Type of offering	Implementation	Maintenance
Standard product	Installation of standard product	Product support
Flexible/local services	Integration of customized solution	Solution lifecycle management

Previously, when a service was realized via a standard product, only support was needed. Today you need to insure that the whole solution is maintained, even when new software releases of individual components arrive or the network changes. Table 8.1 illustrates this shift of conception. This implies a clear need for design and deployment processes that allow for a solid feedback channel from local deployment groups all the way back to design organizations to adopt and anticipate the many customizations that are likely to be incorporated into a product as it is delivered and maintained in the future as a complete solution.

8.2.6   The complete solution

Beyond providing a platform or solution, it is imperative to satisfy all solution needs and anticipate the issues that will arise further down the chain when implementing a solution. Extend the scope of your planning beyond the pure functionality of the solution and look at wider issues such as:

- Operation and maintenance – how will monitoring and maintenance functions interact between operators, suppliers and other players such as aggregators and hosting centers? How does one maintain awareness of end-to-end service performance? Will monitoring of alarms be integrated into the existing NOC (network operations center) or will the solution have standalone systems and maintenance?
- Customer support – likewise, how do the various support units interact and effectively troubleshoot problems?
- Service provisioning and charging – how will subscriber information updates be handled? What charging models are required? Are they feasible?
- Service roaming and interoperability issues – will end-users in different networks be able to use the same service? MMS did not take off until two or more operators in the same market offered the service.

- Reporting and accounting – are all parties in a value chain kept informed about the end-to-end performance of the service? Is it easy to deduce all parties' revenue shares and will clearing of revenues be made quickly and effectively? Have you considered the requirements for data mining?
- Customer training – it is important to consider training for not only the technical receiver organizations but also for sales and marketing groups. Maintaining a strong contact and dialog with these groups ensures that a solution can be competently sold and promoted to consumers.

At the end of the day, one must allow receiving organizations to take their time to digest and absorb a proposed solution design – even after joint planning. Set aside time for this final step of the process and remain open for discussion in order to secure a firm commitment before moving into the next steps of deployment.

After agreement, expect changes along the way, so be sure to negotiate a change management process, establish an approval board and stick to it on *both* sides!

8.3   SOLUTION INTEGRATION

As full-scale integration begins, one must again look carefully over preparations and think carefully about the resources and tools required to execute a successful project.

8.3.1   Team composition and processes

When building a team one should consider not only the core technical competence but also ensure that you have strong project management and quality management competence to manage day-to-day logistics and adherence to processes which should be clearly defined before the project starts. Operator processes can appear at first glance to be overly exhaustive, cautious and may delay the integration, but they are in place to protect the network and revenue stream.

Integration teams should involve resources from all stakeholders to ensure a smooth operation with the same lessons from trials being applied here – stakeholder commitment and resource dedication are key to success. In a more practical sense some of the following tips can be of use:

- Create a responsibility matrix to establish roles and responsibilities and distribute 'cheat lists' of key contacts, phone numbers, daily schedules as well as secondary contacts. In addition, detailed lists should also be maintained with all the key technical information needed – network layouts, IP addresses, firewall configurations, software versions, etc.
- Do not blindly rely on e-mail as an efficient form of communication amongst large teams acting on a tight time schedule – people will either spend too much time reading all the e-mails being sent, or they will simply ignore them. This defeats the whole purpose of communication. Look into more effective methods of communication as well as document-sharing tools.

- Make efficient use of daily update meetings and face-to-face contact to raise and solve issues on the spot both within a team but also with other stakeholders and customers involved in the project.

By reducing the complexity of a solution, one will benefit strongly from a smaller integration team with fewer dependencies – this can make day-to-day activities during the integration far more efficient.

Consider the effects of different time zones in an international deployment where you find yourself making integrations on one side of the world while your support organization is fast asleep in another time zone. By ensuring that a delivery group has adequate support from design teams during work hours, one can ensure quick and informed decisions in stressful situations.

In addition, it is not unusual to get caught up by local laws, so make sure 'simple' things such as visas, drivers licenses and other necessary permits are sorted out well in advance. Make sure your delivery team can effectively communicate in an overseas environment – language barriers not to mention cultural understanding can have a major impact on the stability of a project.

8.3.2   Integration experiences

There are several notorious aspects that become clear after talking to various experienced campaigners that one should be avoided if possible. To start with, let us take a look at some detailed aspects related to the deployment of two media-rich solutions. It is clear from these examples that it is the 'little things' that, if overlooked, will cause many delays.

8.3.2.1   Mobile TV services

An ideal case to illustrate the convergence between the media, IT and telecoms industry is that of the implementation of a mobile TV service. In this case we need to consider the flow of video content from camera to mobile device and the many intermediate steps in between. Figure 8.4 provides a rough illustration of the major aspects of a mobile TV solution. As one moves through this chain, important deployment aspects appear.

When looking at the area of video feeds and encoding, several questions should be considered. What is the nature of the incoming video feed? Is this simply the regular broadcast TV channel feed or is it tailored content for use in the mobile arena. Will the TV signal be received via antenna or is a direct coupling to the TV studio available? What is the aspect-ratio of the video feed?

In the case that the video feed is analog (A/V), then one must ensure that the feeds are separated into video as well as left and right channel audio channels since numerous encoding systems require completely discrete feeds. Many encoding systems run on PC-based equipment with standard video encoding expansion cards. For reasons of redundancy, it is thus common to employ the use of several parallel encoding systems in case of the failure of an encoder for one reason or another.

Simply rebroadcasting standard TV content could lead to problems further down the line. Various TV programs employ the use of overlay text, split screen visuals and other

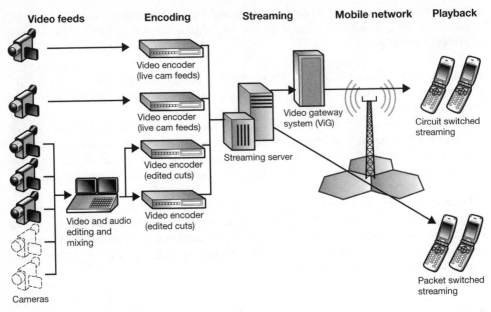

Figure 8.4 A high-level architecture of a mobile TV solution.

content that is very hard to see on a small mobile device, but it also affects the efficiency of the MPEG4 compression schemes used.

Case study: Mobile news TV services

Consider a mobile TV headline news service. Video compression works best with slow-moving or static objects. In this case the subject matter is static 'talking heads' for much of the time as news presenters talk through the latest current affairs. Bandwidth consumption should thus be kept to a minimum and the perceived performance by the end user should be rather good since the compression engine is at its most efficient.

However, on all modern news programs there are generally scrolling text and overlays that convey latest news updates, share price tickers or the names of the news presenters. Encoding this additional content into a mobile feed will result in a reduction in compression efficiency since the encoding mechanism must continually track these continually changing overlays. On a mobile display this information would in any case be too small to read. After compression, however, things are worse, with a lot of distortion at the bottom and sides of the screen – rendering the text even more unreadable since the compression algorithms generally cannot keep up with these changes and the overall perception of the video is greatly reduced.

The choice of video feed is thus very important for use in the mobile environment and it is preferable if at all possible to acquire a relatively pure video feed that has not been overly adapted for traditional TV broadcast. This is especially relevant for nonlive broadcast of video clips, where information in the clip can be displayed before the video clip is selected. See Chapter 4 for more information on the design of interactive mobile TV services as well as a look at video content creation.

From an integration and cost limitation perspective it is always best to locate the encoding functions as close as possible to the video feed locations. The output of the encoders is a compressed, and thus much smaller, content stream. This means that transmission costs are limited when sending the video from the feed location to the relevant mobile network streaming nodes.

Depending on the type of content, the encoding profile makes a large difference. In the case of prerecorded nonlive content, the subject matter is known (static news broadcast, fast sports clips, movie trailers with many scene transitions) and this means that the encoder settings can be adjusted until the output is 'subjectively' acceptable through an iterative encoding process. This is a more optimized activity, but one which also takes more time.

In the case of a live TV stream, the encoder settings are fixed and the output cannot be edited for further improvement. The subject matter is not known in most cases, so the obligatory use of a generic encoder will invariably result in generally poorer results. Also, make sure the encoder settings are set correctly to handle the right format of incoming video – PAL or NTSC.

Looking further into the mobile network and streaming servers, one of the most important aspects is to ensure that the correct encoding format and bit-rate are used depending on the wireless throughput capacity of the network. The key here is to play it safe even for those mobile networks that do guarantee QoS.

For example, in GPRS with a potential downlink bandwidth of up to 50 kbps, one should encode at around 20 kbps with H.263 and AMR-WB codecs for noninterrupted live streaming. In WCDMA networks an average value of 56–58 kbps is used; however, some 'premium' content that requires higher quality encoding can be made at about 120 kbps. Variable bit rate encoders are normally used so these average values are used in the encoder settings, enabling the actual bit rates to rise slightly at times – so it is safe to keep some margin with regards to the standard bearers in WCDMA networks (64 and 128 kbps). Higher encoding rates push the limits of current mobile devices to process incoming video streams.

It is likely that volume-based charging is not applied for the consumption of video content. Instead, one-time fees or monthly subscriptions cover the cost of such a service. In this case it is necessary to ensure that volume-based charging is disabled within the mobile network for streaming services through either the use of a separate network APN, which requires specific device configuration, or by relying on flexible bearer charging – a function of more advanced mobile networks which allows the network to adapt the billing of traffic based on the overall service's charging profile.

On the device side, thorough testing is required to ensure that the device supports each format and bit rate of video content that it will receive. Read more later in this chapter for tips on device verification.

In the end-to-end perspective, make sure network bandwidth allocation and session time-out values are set and the overall network architecture is configured for the carriage of streaming traffic – this means making sure that NATs, proxies and firewalls are properly configured:

- Make sure RTP (real-time transport protocol) and RTSP (real-time streaming protocol) ports are open (554 and 8554);

- Make sure appropriate UDP ports are open;
- Make sure NAT settings are done properly so as not to block traffic in either direction.

The added complexity of integrating solutions across the media and telecoms industry is quite clear.

8.3.2.2 Ringback tones

Ringback tones are a popular service that allow subscribers to further personalize their phone settings. Subscribers can purchase melodies and configure their phone account to play them to incoming callers.

On the surface, this would seem a relatively simple service to roll out – just play back a song instead of a normal beeping tone before the call is answered. Indeed there is no DRM since there is no download of content and all mobile devices naturally support this service. Integration into the IN and call handling functions of a network is, however, extremely complex and touches upon well-established and mission-critical network nodes, i.e. components that control an operator's main source of revenue – voice calls.

As mentioned earlier, one should avoid multiple integration points if possible. In this case this is unfortunately unavoidable. In GSM networks, for example, in a worst case one may have to interact with the following systems:

- MSC – Mobile Switching Center, a key switch in the mobile core network;
- HLR – Home Location Register, manages all subscriber data and the services that a subscriber is registered for;
- Billing systems – required to execute monthly subscription charges and event-based charges relating to the service for both prepaid and post-paid subscribers as well as for content provider settlement;
- OSS – Operational Support Systems, network element management, maintenance and control;
- Customer administration systems – required for service provisioning and subscriber handling;
- WAP/Web/IVR portal – integration required to provide access for subscribers to purchase and change ringback tones;
- SMS-C – optional integration in the case of subscriber purchases and subscription handling being made via SMS;
- Content management systems – required for content providers to upload new material and to manage the existing content portfolio as well as disposal of outdated material.

Suddenly it is clear that this is not such a simple integration after all.

One must then think carefully about the various scenarios and use cases that must be accommodated. In addition to the obvious use cases, there are many other less obvious cases that must be handled according to stakeholder preference:

- How should diverted calls be handled when the new target number may or may not have a ringback subscription? What happens when a call is diverted to voice mail – does the ringback tone continue during the diversion? What if the user is busy or not reachable?

- Should incoming data calls be played a ringback tone? Likewise for incoming Fax calls? Do video telephony calls work as well?
- How are international roaming scenarios handled?
- How does the activation of lawful interception affect the ringback tone service – end users should certainly not perceive any disruption.

Additional aspects relate to the adaptation of the actual content. It goes without saying that the audio clips must be carefully chosen. A short and recognizable clip of the song must be used, but also one that sounds acceptable with faint ringing beeps overlaid.

Historically, one problem had been that each song had to have the beeps manually mixed into the master recording; however, this has been rectified in later solutions. One must, however, still accommodate various different ringing tones for the different countries of deployment.

Finally, even though the used content is an abridged version of a full song, music labels will need assurances that the large volume of content is stored securely in an authorized facility.

8.3.3   General integration aspects

The above examples highlight many specific areas of concern that can hinder integration efforts. The following section addresses common aspects that one should consider.

8.3.3.1   High-level APIs – bell heads and web heads

As we have seen, there is a wide variety of interfaces and technologies – ranging from 'standards' to proprietary implementations. Amongst platform vendors this can become a domain for differentiation, especially during early launches of a technology. Protocols and interfaces generally settle and become consistent as technologies mature and become standard, but never assume that all standards are followed 100 %. Flexibility and research are important to ensure that a solution is readily adaptable – every target environment is different.

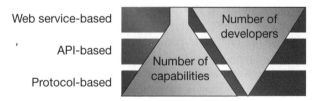

Web service-based

API-based

Protocol-based

Number of developers

Number of capabilities

Figure 8.5 It is easier to find competence to implement high-level interfaces.

Strong network architecture with practical high-level APIs will generally pay off in terms of simplified integration. When integrating towards complicated systems such as charging and provisioning nodes, the value of high-level interfaces becomes clear. While the breadth of features may be somewhat restricted by a more abstracted API, the competence needed to implement and integrate support for a specific interface becomes far more readily available. Contrast the competence and experience that is required to implement a wireless charging transaction using low-level Diameter protocols as opposed to an abstracted high level web services-based function call.

This is an extremely important concept as the telecommunications world moves into, and relies more and more on IT standards technology. It is not feasible to expect media partners, content providers and other solution developers to learn the intricacies of traditional telecoms protocols. By abstracting the protocol complexities of an interface behind a web services module or a Java class library object, one makes development, integration and verification of a feature far more realistic. Keep bell heads and web heads in their speciality areas!

8.3.3.2   Implementing charging functions

Premium-SMS is by far the most popular method used to charge for services in the mobile arena today. Consumers understand the concept, it is simple to use and it is a relatively simple process to integrate for a service provider.

This is important when one compares SMS-C integration to integration directly towards a charging node. The latter involves delving deep into the lower layers of charging protocols where one must program complex functionality to execute a charging transaction, including reservation of funds, charging data record (CDR) creation and exchange, various charging models as well as potentially different nodes for prepaid and post-paid subscribers, etc.

However, as services evolve, more sophisticated business models appear and the pay-per-event model of SMS becomes inappropriate when one wants subscription-based charging or when the price for a service gets higher.

8.3.3.3   Operation and maintenance/reporting

This is often the most forgotten aspect of a solution. A product may have all the whiz-bang features but is rendered useless without an effective O&M interface by which to manage these features.

Systems must also produce useful counters and statistics that not only help diagnose the performance and availability of the node, but can also help to give insights into the end-to-end performance of the overall service that the solution supports. This is of great importance when diagnosing faults and bottlenecks within a system. Despite the fact that all nodes in a solution are up and running, the overall performance of a service may suffer due to poor integration between nodes as well as other unexpected issues such as high network load, software licence capacity limitations, etc.

Thus, important KPIs (key performance indicators) should be carefully defined as a part of the deployment project and these measurements should be readily visible.

8.3.3.4   Customer support

Good support can often make a bad product seem acceptable – it is all about perception. Unfortunately, the opposite is never the case. Poor support will cripple the relationship with a customer despite the best credentials and capabilities of a product.

Figure 8.6 Support is your operational face to the customer.

It is all too often forgotten that support represents your operational face to the customer, both for service providers towards consumers as well for subsuppliers further down the chain. Ensure rapid feedback and engagement in support issues and establish strong processes to handle, track and resolve requests.

8.3.3.5 Logistics

While tackling the many integration issues, the supply of hardware equipment, software and licences must be well scheduled and particular attention should be made to survey the target environment.

All too many common issues can frustratingly delay installations and each wasted day quickly adds up. Beware of common pitfalls such as:

- Where will equipment be installed? Is the correct mains power (and backup power) available? 120 or 240 V/+24 or −48 V DC? Equipment varies and suitable supplies are not always available. It is amazing to see how many times equipment is supplied with the wrong power cables.

Figure 8.7 Logistics: every wasted hour adds up.

- Incorrect versions of hardware and operating systems for application software lead to new costs and installation time. This can be something as simple as insuring the correct hardware model or insuring that the equipment is ordered with hard drives installed.
- Wrong software versions or drivers lead to time-consuming patches and updates.
- Incorrectly licensed capacity potentially leads to capacity issues – often only found well after launch
- Incorrect router, VPN and firewall settings that block specific traffic or ports are extremely hard to trace.

Just as people need visas, so too must equipment deliveries be well prepared. One classic case was that of equipment delivered overseas that was held in Customs since it was not labeled properly when sent. Importing equipment (even laptop PCs) into countries can be a bureaucratic process, where mistakes inevitably result in expensive delivery resources twiddling their thumbs while waiting for their equipment to be cleared.

8.3.3.6 Maintenance windows

When it comes to deployment in a live network environment one must be well aware of maintenance windows within operational systems. Installations must be able to be broken down into manageable blocks to fit in with the only available time slots – often overnight, for example, between 11 p.m. and 6 a.m. In this time it must be possible to install and verify necessary components without extended disruption to the network afterwards. Installation processes that take longer or cannot be broken down will often violate the network maintenance rules and will not be popular.

One must also anticipate lock-down periods where a system is not to be disturbed or modified at all. A Christmas lock-down is a well-accepted concept to ensure that a network remains stable over the busiest period in the year – try not to plan major system installations anywhere near this time of the year.

8.4   VERIFICATION

Testing comes in all shapes and sizes – everything from functional testing and quality assurance to verification of games and e2e system performance tests.

The introduction of new multimedia components such as streaming video, downloadable music and images together with new concepts such as DRM and SMIL (synchronized multimedia integration language) has created an explosion in complexity in the verification and acceptance of new services in the deployment phase.

In this section we will concentrate on verification of components at the application layer as opposed to lower level systems in the mobile network domain.

8.4.1   Device verification

It is important to start the discussion on content-related verification with a focus on the importance on strong device testing. Despite the immense efforts to standardize features such as WAP, MMS, streaming, etc., there is still significant diversity in the interpretations of industry standards by device vendors. Indeed, implementations of such features can vary greatly even between different device families of the same vendor (e.g. between Nokia Series 40 and Series 60).

In addition, one must contend with the natural variance of devices in terms of screen size, resolution and color depth, audio features and input capabilities, as well as support for various media content types. Categorization of devices must go beyond the specified properties of a device. Knowledge of a device's screen size, for example, must be extended to understand the maximum image size that the device can display in its WAP browser, MMS player, photo gallery viewer as well as for use as the phone's desktop. In addition, device CPU power and available memory are a significant concern, especially for demanding functions such as streaming video playback as well as parallel operations.

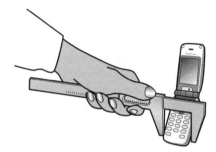

Figure 8.8 Be harsh when verifying your products, or your customers will be harsh to you.

8.4.1.1   Important areas of device classification

The breadth of features one must be aware of is wide, with some specific aspects listed below:

- Image size, resolution and color depth – does the device fully support SVG (scalable vector graphics) and animated GIFs as well as the multitude of other formats? What image formats work best with the phone with regards to both perceived quality and device rendering times.
- Audio content – what is the support for newer audio codec formats such as AAC and AAC+? What encoding bit-rates are supported and does this match the download speed capabilities of the device for streaming?
- Video content – what specific codecs are supported? Can the device practically handle higher resolution video clips with high frame rates?
- DRM support – what specific models of DRM does the device support? Can it be used for more sophisticated business models that require time-based rights access such as monthly subscriptions to downloadable music content? How does the device practically handle the use of DRM 'view once' features used to allow previewing of content? What is the end user experience of using this feature and does content need to be manually deleted after viewing?
- SMS content – does the device support the applicable international character set as well as handle embedded binary content such as simple ringtones and logos?
- MMS content – how does the device support standard MMS message types as well as combinations of content and multiple slide based massages? What are the size limits

for incoming messages, the maximum number of slides per message and the amount of content that fits on the screen of the device for each slide? Can the user easily scroll each slide and save embedded content?

- WAP content – in addition to support for specific browser languages, what are the usable screen dimensions for images and text content? Are soft key prompts displayed on-screen or only accessible via a menu? How wide can page headers be without being truncated? How fast does the browser render content?
- J2ME/BREW/Open OS environment – the exposure of specific functions and features present in the device is governed by the APIs implemented in the device. For example, in the J2ME environment, the implementation of JSRs which relate to specific capabilities varies greatly between phone models. One must check closely the support for specific JSRs as well as the actual versions of the JSR that are implemented. This applies equally in the area of BREW and also in open operating system environments such as Symbian.

As a further burden, it is clear that very subtle bugs and undocumented 'features' of mobile devices easily ruin the end-user's experience. These are the hardest things to find and relate to combinations of aspects that can cause a test to fail. For example, a device specification may support the use of MP3 files as ringtones as well as support the handling of OMA DRM protected content. However it was found in one case that, when a DRM-protected MP3 file larger than 300 kB was downloaded to a phone, the size of the file caused a failure. In addition, if the file had a space in the filename, download of the DRM encoded content would also fail. No mention is made in device specification of these 'limits', and it is here that the majority of problems arise. It is important then, that one tests the performance of new devices within a genuine application environment with live content.

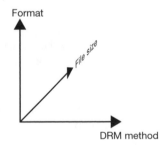

Figure 8.9 Three dimensions of practical content verification.

For this reason, it is crucial that, when verifying new devices, one thoroughly tests the full capabilities of a mobile device to establish its full capabilities as well as retaining knowledge about these aspects when it comes to testing the various content items of a service towards that device. Device capability databases are now a significant asset that many solution providers can charge a hefty sum for access to.

8.4.1.2 Anticipating costs for device verification in the content and application sphere

Typically, one can expect the cost in terms of man-hours to be between 1.5 and 2 man-days of work to verify and classify the content and application related features of a brand new device. For those devices that are an evolution of a well-known series, one should reckon with at least half a day of testing.

There are of course limitless numbers of possible test permutations but it is important to focus on the most important devices. Here we see the benefit of limiting the scope to those devices that are well recognized in a target market as well as to the devices that are in the hands of target user segment. It is of limited value for example to test 3D games for high-end smart phone models if no young gamers have these phones. A greater challenge exists in those regions where there is limited or no operator control over the devices that are released in a market. In this case, one must be well aware of the best selling devices that have the deepest penetration.

Summarizing, device verification is used both to fully classify the properties of a device for future content compliance testing as well as for regression testing of the current content and features within a mobile service.

8.4.2   Content and application verification

We have seen the obvious needs to ensure that content conforms not only to standard but also to the specific properties and quirks of devices targeted for the service. Such testing is closely related to the aforementioned device verification; however, it should be noted that significant gains can be made through the use of automated toolsets to verify the functional and cosmetic conformance of content in multiple devices. Such tools can automatically sift through content items and highlight pure incompatibilities as well as more cosmetic issues for browser-based content such as page titles that are too long to display on a specific device, as well as line spacing and character set support issues.

It is important to take into account the context in which content will be used. As described earlier, the verification of music content for use as a ringback tone is greatly different to the tests required for music track downloads. This implies that there is no standard test specification for content items – it is very dependent on the context of use and it is important to limit the scope of testing to cover just these aspects. It is also important to limit the time and resources required to execute verification in order to address only the key areas of importance.

In an operational sense, as partners successfully integrate their solutions and enter smooth operations, one can rely more on random content checks and be better assured of a content partner's ability to provide quality content that will not cause large-scale incompatibilities. In such a case it is possible to then rely on customer support to catch and rectify the few problems that do arise. Naturally, one must be careful to maintain a strong dialog around the introduction of new devices into a service portfolio as well as keeping a close eye on the volume of customer support requests and complaints that arise over time.

8.4.2.1   WAP content verification

The finer touch in the area of WAP verification involves more of the cosmetic aspects of such a service. While one must check whether a browser has the ability to display your content, the look and feel of the services as well as the user friendliness of its browser can often leave a lot to be desired. Refer back to Chapter 4 for more details on usability tips with regards to WAP application design.

Important areas to manually double-check during integration are:

- Ease of navigation of the service within the overall portal – can the user easily go back into the main portal as well as easily navigating around the service.
- Appropriate rendering of content – check for misalignment of images and text as well as truncation of titles, body text or softkey text.

8.4.2.2 MMS content verification

MMS is somewhat simpler than WAP to verify owing to the relative simplicity of MMS content specifications. Key aspects to check here are:

- Message size – can the device and network handle it? Both MMSC infrastructure and WAP gateways can impose limits on total message size.
- Slideshows – there is varying support for how slideshow content is handled on different devices. For example, on one device model it was possible to play a single music track across multiple slides whereas, on another, developers were obliged to create separate music clips for each slide.
- Slide length – consider carefully the use of slide content that is longer than the screen size of the device, thus obliging the user to scroll through it. In this case, the user must both know to scroll downwards as well as have the time to scroll before the message moves onto the next slide. Optimally, longer content should be spread across multiple slides, thus enabling the user to simply watch the message without interaction.

8.4.2.3 Application verification

Potentially the most complex issue of all is addressing the needs of verification for applications that run within a Java or BREW environment as well as in open operating systems such as Symbian and Windows Mobile. Applications in this area are naturally far more varied and can also pose a potential security threat in terms of malicious use of messaging or other network features as well as accessing personal information from within the phone's address books, calendar, e-mail, etc. A nightmare scenario for operators is also the potential for such applications to swamp mobile networks with spam and other undesirable network attacks.

Case study: Viruses illustrate the need for strong verification

At the time of writing, there are over 10 Symbian OS-based viruses with numerous variants that have the capability to cause damage to mobile devices. This began with the Cabir virus – a proof of concept to prove that even Bluetooth technology was not invulnerable to viruses. Moving on from infection via Bluetooth, the CommWarrior virus spreads itself using Bluetooth as well as MMS, and will reset the infected phone on the 14th of every month.

Similar to the email world, the user is delivered an MMS with enticing subject and body fields such as:

Subject: Norton AntiVirus

Message: Released now for mobile, install it!

Subject: Happy Birthday!

Message: Happy Birthday! It is present for you!

Certainly one of the most dangerous is:

Subject: Virtual SEX

Message: Virtual SEX mobile engine from Russian hackers!

Once a user installs the attached application, their phone becomes infected and will further spread the worm.

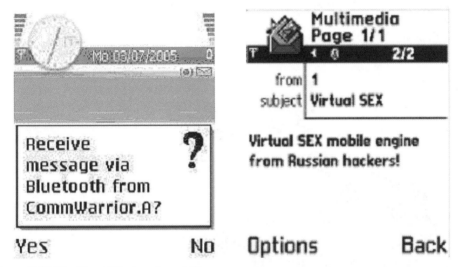

Figure 8.10 The CommWarrior virus strikes.

The need for strong verification is clear and the industry has indeed stepped up to the challenge and provided clear backing for various initiatives to ensure quality in mobile applications. The following initiatives are worth mentioning:

- J2ME, Java Verified – a vendor- and operator-backed initiative to provide industry-wide test requirements and test services for verification of J2ME applications.
- BREW, True Brew – Qualcomm-driven certification scheme that all BREW applications must pass through in order to enter circulation.
- Symbian, Symbian Signed – driven by Symbian, the program promotes best practice in designing applications to run on Symbian OS phones.
- Windows Mobile, Designed for Windows Mobile – Microsoft has also extended their certification schemes to cover certification of applications in the mobile sphere.

Such certification programs are increasingly seen as a quality guage by operators and aggregators and can be a strong marketing asset. In addition to safeguarding against security threats, these schemes also ensure that applications are well structured, have correct spelling and allow the user to pause the application. They importantly ensure that the application can allow incoming telephone calls, messaging, Bluetooth and infrared connections to be presented to the end user.

The costs of taking an application through such testing can be large, since costs increase for each additional phone one needs to test an application on. These costs alone can threaten the slim margins that many application developers have; however, with the industry desperate to engage new users, poor quality or insecure applications present an even larger threat to the reputation of the mobile data industry.

8.4.3 End-to-end testing

At a final stage, verifying applications in live conditions (as closely as possible) will always bring new issues to the surface. It is here that effects of the interaction of different network and solution elements are at last clearly seen.

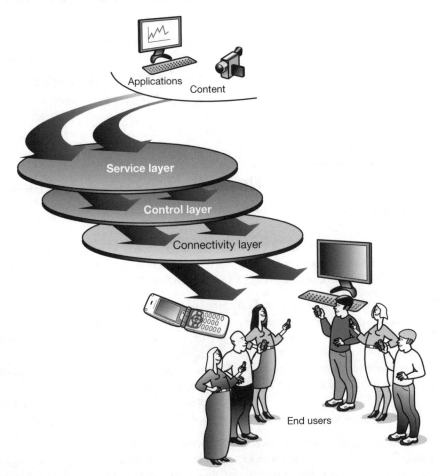

Figure 8.11 End-to-end testing encompasses all elements in the chain.

Take for example the implementation of a mobile e-mail solution. In a live environment, one will see the performance of the solution running together with VPN software on a loaded mobile network together with real devices with limited memory and storage capacity. In such a case, good planning at the development stage should pay off with good performance. Poor awareness of live conditions could result an expensive return to the drawing board.

This is also a chance to conduct realistic load and scalability tests to ensure that the solution itself and also other network elements are able to handle the increased traffic and new functionality that a new solution will hopefully bring.

Implementation of practical aspects such as O&M systems as well as support processes should be thoroughly tested to ensure they can handle the inevitable queries and requests that will arise immediately after launch.

8.5 SUMMARY

Preparation is key; by understanding the target market and agreeing on the scope of a delivery, you can insure a strong start to a deployment project. Be ready for challenges, hiccups, delays and changes of requirements. Strong communication amongst all stakeholders is the key to overcoming these challenges.

The mobile industry, with its huge diversity and rapid evolution, demands a strong and committed approach to verification. Mobile customers expect quality and convenience. By rushing through the verification stage, you will only further damage the perception of mobile data services by providing solutions that either fail to perform or do not provide the quality that is expected of them.

9

Commercial Launch Experiences

At last we approach the day when our application is set loose on an unsuspecting market. No matter how innovative the original concept was or how well the application has been designed and tested, it is almost always the case that the success (or otherwise) of a product or application is determined largely on how well it is marketed.

It might be tempting to launch an application as soon as development and integration are complete, and often there is considerable competitive pressure to do so, but if the initial customer experience of the application is poor, with all the publicity surrounding it, its reputation will be damaged from the start, initial growth will be disappointing and the money spent on promotion wasted.

The period following the launch of a new application is the most critical, for various reasons: it is generally when marketing spend is at its highest; it is when the first, 'innovator' customers evaluate the application; and it is when the most media attention is received. A newly launched application must therefore perform at its *best* on the day of launch. It is not acceptable to sort out an application's teething problems after launch, effectively using customers to test the product. So we reiterate the message of the last chapter that a prelaunch trial must be performed.

Also, remember that, although for you and your team this may be the end of a long and tiring project, from the customer's perspective this is just the beginning. This is your baby and it is your responsibility to see that it survives in the big wide world.

9.1 LAUNCH STRATEGIES

9.1.1 What is the goal of the launch?

When we ask this question to operators and application developers, the answer is invariably 'to generate revenue'. This is understandable – the bottom line is what drives all companies – but there are many indirect effects of launching new applications, and while

Mobile Media and Applications – From Concept to Cash: Successful Service Creation and Launch   Christoffer Andersson,
Daniel Freeman, Ian James, Andy Johnston, Staffan Ljung
© 2006 John Wiley & Sons, Ltd

they may not always be easy to quantify in financial terms, they can nevertheless be of significant benefit to the company. There may be a bigger and better goal for applications than simple direct revenue. Let us look at a few examples.

9.1.1.1 To retain customers

Applications can reduce the likelihood of customers leaving an operator. For example, if a customer has a mobile e-mail address with the operator, changing to another operator would mean changing e-mail address, with all the inconvenience of informing everyone of the change.

From a financial point of view an application that helps to retain a customer secures that customer's voice revenue and avoids the cost of acquiring a replacement customer. These should be factored into the business case for the application as a 'cost of not launching' scenario.

9.1.1.2 To drive voice revenues

Voice traffic is, and will doubtless always be, an operator's principal source of revenue, so if a new application attracts new customers the added voice traffic alone could justify launching the application. A case study from Japan illustrates this well.

Case study: Sha-mail – driving voice through differentiation
At the end of 2000 Japanese mobile operator J-Phone (now Vodafone KK) launched a mobile picture messaging service called Sha-mail (literally picture mail), which included the world's first camera phone.

The phone was heavily subsidized, and 18 months after launch J-Phone had sold 6 million camera phones – 40 % of its subscriber base. The cost of sending a message was very low, at 8 Yen (about US $7¢). The operator indicated at the time that customers were sending on average two to three picture messages per week, so each user contributed under a dollar per month. So was this really a success?

Well yes it was, indirectly. To understand why we have to look back to the situation in Japan in 2000. At the time the three mobile operators in Japan used different mobile network standards, so each operator's handsets only worked on that operator's network. J-Phone was the only operator selling camera phones, and enjoyed this monopoly for 12 months until NTT DoCoMo and KDDI launched competing products. People who wanted to buy camera phones were therefore obliged to become J-Phone subscribers. As a result J-Phone's share of new subscribers increased from 15 to 30 %. Each new subscriber brought added voice revenues, so the real benefit of Sha-mail was not revenues from picture messaging at all, but new *voice* revenues due to the increased market share. It is estimated that voice revenues from new J-Phone customers attracted by Sha-mail were five times the total revenues from picture messages.

It was J-Phone's deliberate strategy to use Sha-mail to acquire new customers, which explains the heavy handset subsidy and the low cost of sending a picture message. J-Phone knew that they had limited time to achieve volume sales before their competitors caught up, so they kept the price barrier very low in order to attract as many customers as possible. If J-Phone had focused on maximizing direct revenues from picture messaging, they may have missed the bigger opportunity.

9.1.1.3   To attract high-end customers

Another valuable outcome of the Sha-mail launch was that it attracted *high-spending* customers from its competitors. Sha-mail users spent US $70 per month compared with US $46 for non-Sha-mail users. The difference was not revenue from picture messages, as we have seen; these customers were already high-spending voice subscribers.

This is likely to be true for all new applications that use the features of a high-end handset. People who are attracted by, say, mobile music or video telephony, are likely to be early adopters and high-spending customers, so these applications attract the highest value customers.

9.1.1.4   To sell more Coke?

The broad penetration of mobile phones, particularly among young consumers, offers endless opportunities for indirect revenue generation. For example, we have already discussed SMS being used in TV programs for voting and interactive chat shows. The SMS traffic generates income for the operator and the application provider, but the prime reason is to increase viewing figures for the TV program. Similarly, some film distributors create mobile previews of new movies, which are offered to mobile customers – sometimes free of charge – in order to promote the movie. The principal goal is bigger cinema audiences.

Coca Cola has also been an enthusiastic user of the mobile channel, as this case study illustrates.

Case study: Coca Cola mobile marketing
Coca Cola identified a strong correlation between the users of fun and entertainment mobile services and the target groups for their own marketing. They developed a number of campaigns and associated mobile services with the main purposes of strengthening their brand and selling more soft drinks.

One such campaign offered customers the possibility to send fun voice messages to their friends and enter a competition to win VIP tickets to concerts of high-profile artists like REM and Santana, as well as millions of smaller prizes such as background images, ringtones and MMS pictures.

To enter the customer simply bought a bottle or can of Coke and made a call to a number printed on the label. He/she could then choose to send one of a selection of fun voice messages to a friend, for example:

Good afternoon. Sorry for interrupting, this is the police. We have to make a telephone alcohol test. Bring your mouth close to telephone and blow forcefully into … Oops! Stop! Your levels are unbelievable! Stop blowing and stay where you are! We are coming for you!

During the campaign 29 million Cokes were sold – a small but significant increase from the previous year – and an average of 850 people per day called the campaign phone number. The results of a post-campaign survey showed a positive reaction from the majority of customers with almost two-thirds finding it fun, and around 50% of participants sending a fun message to a friend.

9.2   PACKAGING AND BRANDING

Packaging is a well-tested marketing tool. There is a classic case of a home pregnancy test kit being marketed to two distinct customer groups: those who were trying to become pregnant and those who feared they were pregnant. The package targeted at the first group included a pastel-colored box with a soft-focus picture of a young baby and the tag line 'Planning your family's future'. The box was placed in the store together with the baby care products and priced at around US $35. The package targeted at people who feared they were pregnant was in a plain white box with a picture of a test-tube, a mass of chemical information and the tag line 'Peace of mind in five minutes'. The box was placed in the store together with the condoms and priced at US $30. The contents of both packages were identical. Both products sold well.

As can be seen, even though the same product satisfied both customer groups, the needs of each group were very different. The perceived value of the packages differed between those wanting a positive result and those who hoped for a negative: the US $35 box is an investment in a child's future where only the best is acceptable, whereas the US $30 box was a fair price to pay for peace of mind.

So packaging works. We will now look at ways to package mobile content and applications in order to increase the perceived value of our services.

9.2.1   Branded packages

It is common today for an operator's content and applications portfolio to be packaged under a single brand. Vodafone Live is probably the most prominent, but there are many similar examples: t-zones from T-Mobile, O2 Active, Telia Go, MTNLoaded, SmarTone *iN!* and Orange World.

The brand gives the package (apparent) coherence, makes it easier to market and increases its perceived value. Incorporating content from established, quality brands such as MTV, Disney, Universal and the BBC, as well as recognized local brands can further strengthen the package.

9.2.2   Theme/event packages

Mobile packages can be created on a particular theme or in conjunction with a specific event. These are short-lived but highly effective means of attracting customers' attention and driving revenue.

AT&T Wireless offered a package of mobile services during the 2004 Athens Olympics in conjunction with NBC and implemented by US mobile software company Crisp Wireless. Content included schedules and results from events, video highlights, athlete biographies and medal counts, as well an interactive Fantasy Olympics game where mobile users could compete with each other to win medals. More than one million people used the service during the 17 days of the games.

The advantage of the themed package is that the theme can be chosen to appeal to known high-value customers, or to break into currently underserved customer segments. A package is not restricted to mobile content and applications and can encompass cross-media promotional activities and memorabilia.

Case study: the movie *Sahara*

As part of its collaboration with United International Pictures (UIP), MTN South Africa created an exclusive package to coincide with the release of the movie *Sahara*. Movie-related mobile content, including wallpapers and free video trailers, was provided on MTN's mobile portal MTNLoaded. At the same time the official movie web site featured an MTN competition to win a cell phone and a click-through banner to MTNLoaded (Figure 9.1).

Exclusive screenings of the movie were held for MTNLoaded subscribers in four South African cities; MTN and UIP held a joint media release; and a number of radio and TV interviews were held on programs targeting youth and teenagers. A competition offering exclusive movie memorabilia was also initiated.

During the 6 week campaign period 3790 video trailers were downloaded – a significant number, considering that at this stage a total of 7500 billed videos per month were being downloaded from the portal. Content Manager for MTN South Africa, Thabiet Allie, states: 'This promotion and others like it have demonstrated to us the power of collaborative advertising. It has strengthened MTN as a key player in mobile entertainment and has provided UIP with a direct channel to key customer segments.'

Figure 9.1 Cross-promotion of MTNLoaded and UIP © 2005 Bristol Bay Production.

UIP was the first film distributor in South Africa to use the mobile channel in a marketing campaign. UIP was impressed with the results and intends using it as a key part of its promotion strategy in future movie releases.

9.2.3   Handset packages

One of the best opportunities to introduce mobile content and applications to customers is when they purchase a new handset. By bundling content and services with the handset, the package is perceived to be of greater value, customers are persuaded to purchase a more advanced phone that they otherwise would, and they are much more likely to use the services. It does, however, require some coordination to achieve.

Case study: Oi Brazil
Brazilian mobile operator Oi successfully penetrated the under-20s market by offering handsets customized with content and applications around popular youth themes:

- For the 8- to 12-year-old segment, particularly females, the Oi Xuxa service included a specially designed handset that featured images of Xuxa, the most famous children's TV entertainer in Brazil.
- For teenagers, Oi MTV incorporated a Motorola handset and a music service launched in association with MTV.

Both products featured tailored menus and services designed to appeal to their respective target segments, with a focus on messaging and interactivity. Within 6 months Oi had attracted 182 000 users to Oi Xuxa and 168 000 to Oi MTV.

Prior to launch, Oi studied everything from the color and weight of the handsets to how they would deliver the content. Supporting this was a strong synergy between the operator, handset manufacturers, application and content providers and the chosen services.

9.3   PRICING AND REVENUES

It is strange that such an important topic as pricing is often neglected within the mobile industry. We advise many operators and service providers on their marketing strategies and frequently we discover that there is no team, nor even sometimes any individual, responsible for determining how new services are priced. We find this strange, particularly as the price of a service directly affects the number of people who will use it, the revenues generated and its ultimate profitability.

There are three aspects to pricing we shall consider: pricing *strategies*, pricing *models* and pricing *levels*. We will then consider how to divide revenues between the various members of the mobile value chain.

9.3.1   Pricing strategies

Before we even think about setting price levels, we must think about what the overall goal is for our services and applications. Are we targeting high-end users with premium content or low-end customers with basic SMS services? Does our application require a critical mass of subscribers to take off? Are we offering content in order to promote another service or application? Or are we aiming for an optimum balance between subscriber volume and revenues? The pricing strategy we choose should support the long-term goal for the service.

As we saw in the Sha-mail case, J-Phone's launch strategy was to attract new customers by offering camera phones and a unique picture messaging service. Supporting this was the strategy of setting low prices for handsets and messages. Without this pricing strategy the overall goal would not have been achieved.

In the Philippines SMS was free for up to a year, not actually by choice but due to billing issues, but the result is that now 113 million SMS are sent per day by 37 million mobile users – an equivalent of around 100 messages per user per month.[1] No finance director would approve offering a service free for 12 months, but just look at the consequences. SMS is a perfect illustration of how the full potential of a service is only realized when it has reached mass-market penetration.

The ubiquity of SMS has led to its being used for applications that would have been impossible to foresee 10 years ago. If SMS had been actively launched by operators it is doubtful whether the business case would have included TV voting, mobile marketing, chat and dating, sport alerts, information services, banking authentications, ringtone downloads, and the countless others applications that today use SMS either directly or indirectly. Just think what the opportunities for MMS might be if and when it reaches the same level of penetration. Yet do we see operators setting the price of MMS low enough to encourage mass take-up? Hardly.

We should also remember that as technology develops applications become more sophisticated and can command higher prices, so it is good practice to create a broad base of customers as fast as possible with low-priced applications so that there will be a bigger addressable market when we launch the more advanced stuff.

Much can also be learnt from Internet pricing strategies, where limited-functionality applications are supplied free with the option to purchase an upgrade to the full version. For example, Adobe supplies its PDF Reader free to anyone, but charges those who wish to use Acrobat Writer to create PDF documents. The widespread use of the free Reader creates the demand for Writer, which is where Adobe earns its money.

9.3.2 Pricing models

A pricing model defines how the customer is charged for a service. For example voice calls are charged by time; SMS and MMS are charged per message, and so on. Pricing models have been the cause of much customer confusion, particularly when it comes to data applications, but there is now a general trend towards simpler models.

The golden rule is that people need to be in control of their costs and to understand how they are being charged. Some may prefer to pay a fixed price for a service, in the knowledge that this is all they will pay, rather than risk running up a large bill with invisible incremental costs. Conversely people are reluctant to pay a large upfront fee

---

[1] Source: EMC.

for a new service if they are unsure of its benefit or whether they will use it regularly. Alternative models should thus be available so customers can choose the payment method that suits them.

There is no single pricing model to suit all applications (other than 'free', which would be popular with customers but not very profitable). The most appropriate model for an application is generally the one where the cost of using the application most closely reflects its perceived value. Let us look at several pricing models and the applications for which they are most suited.

9.3.2.1 Time-based pricing

Voice calls have historically been priced according to the duration of the call, and it is generally accepted that the more you talk the more you should pay. In most countries only the sender pays, although in some countries, the USA and India are two examples, both the sender and the receiver may pay.

Comviq in Sweden introduced an interesting model where the operator actually paid its customers to receive calls. The aim was to attract new customers and to encourage people to spend more time talking.

Prior to the introduction of GPRS, mobile data services used circuit-switched connections with time-based charging. This worked out very expensive, particularly as the data rate of 9.6 kbps was slower than a typical fixed modem, and because the poor performance and reliability of early WAP services meant that people were often paying to wait for a response from a nonfunctioning server.

Interestingly a few mobile operators have continued with time-based charging even for GPRS-based browsing, which is normally charged by volume. The logic behind this is that customers are familiar with time-based charging and there is often a direct relationship between the time spent browsing and the value of a WAP session. Many data applications, however, do not use the network connection continuously; e-mail, for instance, loads the network only when an e-mail message is sent or received, and to charge customers for the time spent reading and composing e-mails offline would be perceived as unfair and would probably be prohibitively expensive.

9.3.2.2 Event-based pricing

Event-based pricing is used for normal SMS and MMS messaging. The sender pays a fixed price per message, and the receiver pays nothing. When MMS was introduced one or two operators toyed with the idea of charging people to receive MMS messages, since the receiver initiates a WAP connection to download the messages, but this was very quickly quashed in favor of retaining the same model as SMS, which is simple to understand and provides a clear link between cost and value.

Event-based pricing has also historically been used for premium SMS services, such as news alerts, jokes, horoscopes, etc. The disadvantage of event-based pricing is that every event is priced the same, regardless of the content. This is not such a problem with SMS services when all content consists of short text, but as content becomes richer there becomes a need to set different prices for different content types, which is solved using content-based pricing (see Section 9.3.2.5).

9.3.2.3   Volume-based pricing

Volume-based pricing was hailed as one of the main benefits of GPRS. It is certainly a major improvement over time-based pricing, and it sounds great – you only pay for the data you use. In reality, however, it confuses people more than it helps.

The principle is simple: customers pay for the amount of data transferred, say 1¢ per kilobyte. The trouble is most people have no idea what a kilobyte is, and even if they do, they do not have a feel for how many kilobytes, say, 20 min of WAP surfing will generate. Hence they feel they have no control over their costs. And if people feel they have no control over how much something will cost they will probably not risk using it.

The one advantage of volume-based pricing is that it allows for occasional or trial use of data services without the user committing to an up-front fee. The price per kilobyte must be kept very low so that the user is not shocked with a hefty data bill at the end of the month.

9.3.2.4   Fixed-price bundles

The difficulty of explaining volume-based pricing has led many operators towards fixed-priced bundles. US operators have generally avoided volume-based pricing altogether. Sprint, for instance, offers a range of voice minute bundles with data offered as an option at a fixed monthly fee price for unlimited use. 3 UK offers a range of voice and SMS bundles, with free data browsing and content charged on a pay-per-item basis.

Case study: Orange France – Sans Limite
Orange launched GPRS to the French consumer market in May 2002. Orange had already successfully created a market for data applications (based on WAP over circuit-switched data), and wanted to evolve these to GPRS, but without confusing customers with technology terminology.

Orange realized that the customer experience of data applications centers on the handset, so the launch consisted simply of a new range of handsets and a new tariff scheme called 'Sans Limite' (Without Limit). This offered unlimited access to data services for a flat-rate fee of €6 per month, which applied to both circuit-switched and GPRS access.

From the customer perspective, Orange's offering was merely an extension of its existing portfolio with new phones and a new pricing model. Customers were completely

Table 9.1 Vodafone Mobile Connect business tariffs

Price plan	Price of data card	Monthly fee	Data included (per month)	Additional charge per MB
Mobile Connect 5	£199	£11.75	5 MB	£2.35
Mobile Connect 75	£149	£23.50	75 MB	£1.76
Mobile Connect 200	£135	£35.25	200 MB	£1.18
Mobile Connect 450	£129	£53.00	450 MB	£0.88
Mobile Connect 1000	£99	£83.13	1000 MB	£0.59

Source: Vodafone

unaware of the new GPRS technology and Orange avoided any confusion with volume-based vs time-based pricing.

The Sans Limite model was extremely successful, attracting over half a million new data subscribers within 7 months of its launch. Note that not all of these were GPRS users – many continued to use circuit-switched data, but switched to the fixed-price model due to its simplicity and in order to control their monthly cost.

For business use, operators generally offer a PC data card with a range of data bundles. Vodafone's Mobile Connect offers five options, based on the customer's expected use (Table 9.1).

9.3.2.5 Content-based pricing

This has become the standard for downloadable content such as ringtones, wallpapers, games and videos. It is a natural way of pricing – after all it is the content that people value. Content-based pricing allows each content item to be priced separately according to its perceived value, something that is not possible with other pricing models.

Care must be taken not to be over zealous and set a different price for each and every content item. For instance, it is possible in principle to set a different price for weather, news, sport and jokes, but if there are too many options the customer loses track of the cost of each item and we are back to confused customers. Inevitably there has to be a compromise between flexibility and simplicity.

Then we have the dilemma of whether to charge customers for browsing, ordering and downloading content. A person who buys mobile content from a web site is paying an ISP to connect to Internet, so is it fair to charge customers to browse a WAP site? If a user orders an item of content by sending a keyword via SMS, or by calling an IVR number, should he/she be charged for the SMS or the phone call? And should there be a charge for the WAP session used to download a content item?

Many operators do charge for some or all of this, with the justification that the user is using network resources, but this negates the clarity of content-based pricing. If we say that a ringtone costs $3, it should cost $3, not $3 plus 10¢ for the SMS to order it plus 2¢ to download it.

If we ran a shop we would not charge people to walk around looking at the shelves, we would invite them in to browse freely what we have on offer and only ask them to pay for the goods they buy. The trend is also moving in that direction with mobile content. Operators are increasingly introducing free browsing or fixed-priced all-you-can-eat models, as already described.

9.3.2.6 Subscriptions

Subscriptions are a good way of generating regular income from customers who wish to receive content updates or alerts on a repeated basis. Examples are horoscopes, joke of the day, breaking news, religious and inspirational quotes and daily cartoons.

Subscriptions can be implemented such that the user requests, say, a daily alert, which he receives and pays for each day until further notice; or he pays a monthly fee in advance and receives the alert every day for a month. The former may be more complex to implement for prepaid users, as the account of each subscriber must be checked for funds in real time before the alert is sent. Subscriptions can also be used in a creative way to promote new content.

Case study: WIND try and buy
Italian operator WIND offered its customers the opportunity to subscribe to a range of MMS content push services with the first three MMS messages sent for free. Twenty-five different content services were available; customers could subscribe to as many as they liked and unsubscribe at will. The result was that 41 % of customers were still subscribed to services after 46 days.

9.3.3 Pricing levels

Now we come to a subject on which it is tricky to give advice. This book has five authors who, at the time of writing, are located in Sweden, France, Spain, China and South Africa. If each of us were to quote the going rate for, say, a polyphonic ringtone in our respective countries, we would have five different prices. And hence we have another golden rule:

The prices of content and services are determined by the market.

Or at least they should be. In reality the price of a service is often determined either through semi-educated guesswork, or is based on the cost to the operator or service provider, which generally has no relation to the perceived customer value. Let us take MMS as an example.

The industry perspective
MMS allows our customers to send text, pictures, audio and video; its value is therefore higher than SMS, which is limited to text only. MMS is a natural evolution of SMS and so will be quickly adopted by our customers. The cost of transmitting an MMS can be two to three times higher than that of an SMS, hence the price must be set higher. This is actually a good deal for our customers because the price *per byte* of an MMS is substantially less than an SMS: an MMS with a 20 kb picture is around 14 times the size of an SMS but our customers pay only three to four times the price.

The customer perspective
SMS gives me a simple and cheap way to stay in touch with my friends and family and it has become an integral part of my life. Even though MMS is a richer medium, the circumstances in which I really need to send a picture are few, hence MMS has not achieved the same prominence in my life as SMS, and I therefore do not value it as highly. I will not pay a higher price for a service I perceive to be of lower value. Logically, therefore, I feel an MMS should be priced *lower* than an SMS in order to compensate for its lower value and to encourage me and others to use it more.

Often operators set the initial price of a service high, on the somewhat dubious belief that it is easier to decrease the price than increase it. This may be so, but decreasing a price is also not ideal – it effectively says to your customers 'well the thing we just launched clearly isn't as popular as we thought it would be so we can't make as much money from you as we'd hoped to, so we're reducing the price to a more sensible level.'

And the customers thinks, 'so this thing you've just spent a lot of marketing money telling me how good it is, is now worth less, so clearly it isn't as good as you claimed. Maybe I won't bother.'

It is surely better for all concerned that we set the price at the right level in the first place. However, what is the right level? In Section 9.3.1 we described how a pricing strategy should match the overall strategy for a service, so we might have a number of options: we could can set a low initial price 'for a promotional period' in order to persuade people to try the service; we could can set the price to be the same or lower than an existing service in order to migrate users to the new service (e.g. SMS to MMS); or we could maximize the revenues from as many customers as possible. The latter example requires us to determine the *optimum* price for the service.

9.3.3.1 The optimum price

Figure 9.2 illustrates the relationship between price and volume of sales. As the price of an item increases, the sales volume drops as fewer and fewer people are prepared to purchase it at the increased price. The rate at which the volume drops is called the price sensitivity and varies from item to item. The optimum price is where revenue (price × volume) is maximized.

So how do we determine price sensitivity and the price customers are prepared to pay for different services? Well, obviously we have to ask them! We must organize customer focus groups in order to measure people's interest in our services and applications and how much they would be prepared to pay for them.

Even with this information it is a good idea to offer a limited user trial to measure the actual interest in different services and to confirm pricing levels.

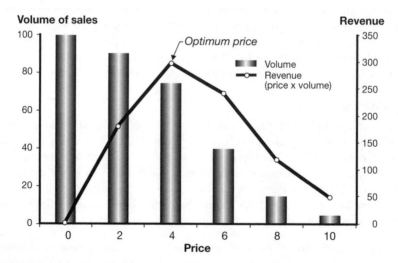

Figure 9.2 Determining the optimum price.

9.3.3.2   What about the content owners?

So far you might have got the impression that the mobile operator is the only one that decides on the prices for mobile content and applications. This is often indeed the case. However, content providers and application developers can also influence the price, especially if there is a revenue-share agreement in place. In fact a content provider often has a good feel for the market price for its content if it is already being supplied to other markets.

All content costs money to produce and sometimes there are license or royalty fees that must be paid. A content provider may require a minimum amount of revenue per item in order to cover development costs and fees, which the operator or service provider must transfer to its customers by setting the appropriate price.

9.3.4   Revenue sharing

Now we have agreed on our prices we can discuss how we are going to split the revenues. As we discussed in Chapter 1, most operators today are prepared to share revenues from content and applications, but there are various ways in which this can be implemented.

9.3.4.1   The revenue chain

There are often several players in a revenue chain that the poor customer has to finance, including the network operator and service provider (sometimes one and the same), a content aggregator, content/application providers and the content creators and application developers. For some content there may in addition be a licensing body that manages content rights, and for music there are publishers, producers, distributors and editors as well as the original artists. So, as you can see, the few dollars paid for a ringtone have to be spread pretty thinly.

9.3.4.2   Revenue share models

There is no standard model for revenue sharing, and the service provider or operator may have different agreements with each supplier. Here are some examples:

- For each content item downloaded, and regardless of price, the service provider pays a fixed percentage, e.g. 50 %, of the net shareable price (i.e. the retail price minus VAT and other taxes).
- The service provider again pays a fixed percentage, but this is negotiated for each price point. For example, the operator may pay 50 % on US $2 items and 55 % on US $5 items. This allows a supplier to negotiate a higher percentage at the price point for which it supplies most content and take a lower percentage for other price points.
- The percentage share follows a sliding scale based on the content value and the volume of downloads, i.e. the more content sold and the higher the value, the higher the percentage received by the supplier.

- For commodity content such as news, weather and financial data, which the service provider may wish to offer free to its customers, the supplier may be paid a fixed monthly fee.
- Subscription services are often discounted from the one-off price, so suppliers may have to compromise on the revenue per item in order to stimulate higher downloads.

Operators rarely share traffic revenues. There seems no logical reason for this other than that they own the network and so consider everything traveling on it as their property.

9.3.4.3   Revenue share percentages

Revenue share percentages are negotiated between the operator or service provider and their content/application suppliers. Rates vary widely and are not often published. i-Mode is an exception, where NTT DoCoMo has from the start advertised the 91 percent it pays its official content providers in order to encourage the development of high-quality i-mode content and applications.

As a rule of thumb, most operators will share between half and two-thirds of content revenues with their content suppliers. They may increase this percentage for premium content where license costs must also be covered.

9.4   PROMOTION STRATEGIES

9.4.1   Promote applications not technology

It should be clear by now that a launch should focus on the applications and the customer benefits they provide, not the technology that lies behind them. WAP is the classic example of a technology launch, but there are also examples where the acronyms GPRS, MMS and 3G have all been the center of promotion campaigns, instead of the services that they enable.

We have seen many a company slide presentation that starts by describing how an application functions, and finishes with the customer benefits. If these presentations were simply given in reverse they would be easier to follow and would create the right mindset in the audience (and the presenter!) that it is the customer that comes first.

However, despite our best intentions, it is not always possible to control the entire launch experience. For instance, some operators deliberately avoided using the term MMS in their launch material and used a phrase such as 'picture messaging', but were powerless when it came to handsets, many of which use 'SMS' and 'MMS' as menu items!

9.4.2   Revolution – the big bang

Mobile applications can be launched with great fanfare or they can be slipped out virtually unannounced. There is no right or wrong way. The scale of the launch is of course determined by how much marketing budget is available; but do not worry, a lack of funds does not prevent highly effective marketing, nor does a big, expensive campaign guarantee rapid take-up. A successful launch is more to do with correctly setting and meeting customers' expectations than raw cash. Let us look first at some high-profile launches.

9.4.2.1 WAP

Despite huge publicity (perhaps because of it), WAP failed to produce the much-heralded mass-market mobile Internet. In fact there was such bad publicity surrounding the poor quality and performance of WAP compared with what was promised, that even if it had functioned reasonably it would have been very un-cool to admit to using WAP, so key early adopters steered well clear of it.

The press and media effectively 'killed' WAP and the public soon forgot it. However today WAP hits have reached an all time high, it remains an integral part of many mobile applications, including MMS, and it provides the download mechanism for much of today's mobile content, such as ringtones, wallpapers, videos and games. The shift from promoting the technology of WAP to the services that it enables has, ironically, led to its widespread use.

9.4.2.2 Vodafone Live

In October 2002 Vodafone launched Vodafone Live in six markets, supported by a £100 million advertising campaign featuring the soccer star David Beckham. Vodafone took a leaf from its newly acquired Japanese subsidiary J-Phone, and took control of the entire mobile value chain from specifying the design of handsets to the selection of content and applications. The launch was an extremely high profile affair and encompassed all traditional and digital media channels.

It is unlikely that Vodafone recouped the cost of this campaign from direct data revenues, but it attracted many new, high-spending subscribers, and it succeeded in its principal aim of strengthening Vodafone as a global brand, strong on innovation. The Vodafone Live launch also made a huge impact within the mobile industry and changed the way many operators packaged and launched mobile applications. It is arguably the best thing that has happened to the industry since i-mode.

9.4.2.3 3G

The hype that has surrounded 3G masks the fact that, technically, 3G really only provides a faster and more efficient network than 2G. Back in 2000 mobile operators, notably in the UK and Germany, spent billions of dollars acquiring 3G licenses, which created high expectations in the media. Fortunately it took several years from the issuing of 3G licenses to the networks going live, in which time much of the hype had subsided and operators could manage customer expectations in a controlled manner.

As a result, the focus of many '3G' launches has been on services such as video telephony and music downloads, with the technology effectively hidden from the customers. In fact mobile provider 3 profiles itself not as an network operator at all but as a mobile media provider. It proudly claims to be the first mobile provider to employ a managing editor for content, and it has introduced the concept of 'Today on 3', more akin to a TV channel.

9.4.3 Evolution – one step at a time

A cheaper, lower-risk and, many would say, more effective launch strategy for mobile services is to take an evolutionary approach. Rather than promote a service as something that is going to change the world and improve the lives of all its inhabitants, we take an incremental approach to launch and focus on specific customers and specific customer benefits.

Even if a major investment has been made in the launch campaign it will be necessary to maintain ongoing marketing activities in order to evolve the portfolio and to maintain customer interest. So revolution and evolution are complementary.

9.4.3.1 Building on existing behavior

People are mostly creatures of habit, resistant to change. A good launch strategy is thus to demonstrate how a new application is a development of an existing service and/or builds on existing customer behavior.

Telenor in Norway promoted MMS as 'a picture with as much text as you like', thus showing the similarities and advantages of MMS over SMS. As a result of this and carefully focused campaigns to persuade customers to purchase and use MMS-capable handsets (such as handset subsidies and regular free MMS campaigns) a very high proportion of Telenor's customer base now use MMS, many using it instead of SMS for text-only messages.

Vodafone actually took this approach with its relaunch of Vodafone Live with 3G. Having already established Vodafone Live as its flagship for consumer content and applications, it promoted its 3G offering as an enhancement to its existing portfolio, focusing on mobile music, video clips and 3D games.

9.4.3.2 Continually adding value

Taking a step-by-step approach to launch means there is always something new on offer. This is important for younger users with limited attention spans. When Kuwait operator Wataniya launched MMS, it introduced a new content category every month. WIND in Italy took a similar approach, with new content ideas continually being added to its portfolio. The added benefit of this approach is that poorly performing content can be removed and replaced with something new. It also gives feedback on the most popular content and gives excellent insight into the sort of customers using the service.

A themed approach is also a very effective way of targeting customers with a common interest as well as making it possible to continuously add value.

Case study: Cineman

In Switzerland there is a movie-based mobile portal called Cineman (www.cineman.ch; wap.cineman.ch). Cineman started with basic information on current and upcoming movies and has gradually expanded to include movie news and previews, screenshots and video trailers. The content is kept constantly up to date and as the portal has expanded both the number of customers and the traffic generated have increased significantly.

9.4.4   Viral marketing

Viral marketing happens when customers recommend a service to others such that self-sustained market growth is achieved. Viral marketing is the cheapest and arguably the most effective form of marketing. It is also the most difficult to control because it lies entirely in the hands – literally in the case of mobile applications – of the customers; it cannot be easily influenced by an operator or service provider.

The success or failure of a movie at the box office is largely determined through viral marketing. The movie distributor will release a prelaunch trailer, issue campaign posters and arrange preview screenings for film critics, but the ultimate number of 'bums-on-seats' relies heavily on the first people to see the movie recommending it to their friends and colleagues. This is a highly visible example of the market adoption curve described in Chapter 2, where early adopters evaluate a new service (in this case a movie) and recommend it (or not) to following customer segments.

The key to good viral marketing is (a) the successful targeting of early adopters, and (b) the ease with which recommendations of the service or application can be spread to others. In the case of mobile applications we are lucky because the mobile phone is itself the perfect viral marketing tool. The initial growth in SMS occurred entirely through viral means. In fact SMS is unique in being both the application and the channel through which it was marketed. However, there is a dilemma brewing in the industry.

9.4.4.1   To share or not to share

Currently mobile users are able to transfer most content items in their phones to another user's handset. Monophonic and polyphonic ringtones, logos, wallpapers, text and picture messages can all be forwarded to other users either via the network, using SMS, MMS or e-mail, or directly between two handsets using infrared or Bluetooth. This is the perfect condition for viral marketing where the barriers to sharing content are low. It is to a great extent how the existing market for mobile content and applications has developed.

The dilemma is that content developers do not earn revenue from forwarded content. Content developers put considerable investment into creating content and ideally they wish to be paid for every customer that uses it. However, if people are prevented from sharing content, many people will remain unaware of the possibility of downloading content for themselves, and the market for content may never develop.

To date most content providers have taken the stance (although they are given little choice) that it is better to allow people to share content and thus develop the market than to try to impose restrictions. In our view this is sensible as the benefits of viral marketing almost certainly outweigh the loss in revenues. It has also been found that users attribute a higher value to content that can be forwarded to friends.

It is also worth remembering that the most downloaded content is ringtones, which people use to personalize their phones. By definition, a personalized phone must be different from other phones, so even though people could obtain their ringtones from other users, in practice most prefer to choose their own tone, so the issue of content sharing is minimized.

However there is one content category that is now forcing the industry to change its stance on content sharing – music.

9.4.4.2   Music and digital rights

Mobile music downloads and 'truetones' (ringtones made from samples of original music tracks) are protected by copyright in the same way as music on any other medium. The music industry, battered by its experience with early peer-to-peer free music services such as Napster, as well as the growth in illegal Internet downloads, insists that it must be impossible for anyone to distribute mobile music content without a license fee being paid to the music label for each copy of the content – including content sent from one user to another. The mobile industry has thus been compelled to introduce digital rights management.

Chapter 6 described the various implementations of DRM as defined by the Open Mobile Alliance. DRM is sometimes portrayed as a means of preventing the distribution of content, but this is not the case. DRM in its full, super distribution, form allows the free distribution of content between suppliers and users and from one user to another, with a license payment being made as required on each content item. Unfortunately, the most commonly used DRM method used today is Forward-lock, which *does* prevent the sharing of content – copyright is protected by disabling the handset's 'send' function, thus preventing mobile users from forwarding content to others.

Forward-lock is a crude way of enforcing rights, but it has at least made it possible to offer real music tracks and ringtones on mobile phones. It does prevent viral marketing of music content, though, which may limit its growth. Add to this the need to pay music royalties, which reduces operator margins, and the fact that Internet download sites such as Apple's iTunes have set the price of a full music track at around $1 per download, and the business case for mobile music starts to look a bit shaky. In fact mobile operators are in the bizarre position of having to offer full music tracks at a lower price than ringtones: in the UK a full music track costs £1.50 whereas a truetone is priced at £3.50!

This is a shame, because mobile and music are perfect partners. We believe the market for mobile music would be better served if the mobile and music industries took a more progressive approach and used the mobile phone as a marketing channel as well as a delivery and playback mechanism. Instead of (or in addition to) offering full music tracks for download, music labels could produce shorter, royalty-free music tracks of new singles that users could download and exchange freely. The mobile track could act as a voucher giving a discount on CD albums or future Internet downloads.

9.4.5   Ongoing promotions

Even after the launch campaign has ended it is critical to maintain regular ongoing marketing activities in order to promote new content and to expand the customer base. A supplier of content and applications may even place a requirement on the operator or service provider to perform a number of promotional activities as part of its content supply agreement.

Promotions come in many forms, just as in any industry. Some ideas are:

- *competitions* – download a Black Eyed Peas ringtone and win two tickets to the concert;
- *special offers* – 50 percent off all US $5 games (one operator saw a 6-fold increase in games downloads from such a campaign);
- *content bundles* – BOGOF (buy one get one free).

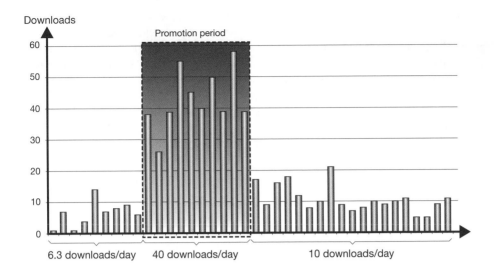

Figure 9.3 The effect of a banner promotion.

9.4.5.1 Simple and effective

If you are skeptical of the effectiveness of promotional spending, Figure 9.3 illustrates the result of a simple web banner placed for a 10-day period on the home page of an operator's mobile portal, advertising a new content category. As can be seen, not only did the banner increase content downloads by a factor of 6 during the campaign, the average number of downloads continued at a higher level than prior to the campaign, due to more customers becoming aware of the service.

9.4.5.2 Mobile marketing

The mobile phone is itself a perfect marketing channel. 3 Italy used MMS as a promotion channel in a campaign called Adesso3. A daily MMS with three (of course) pictures was sent to all customers who had consented to receive commercial communications from the company. The result was that those who experienced Adesso3 consumed twice as much content as those who did not.

Trials of SMS marketing have indicated that an advertiser can get a 10–15 % response rate from an SMS advert. This is much higher than with traditional media, where a response rate of 1 or 2 % is respectable. Add to this the ability to directly target specific individuals on a device permanently in their pockets, and it is clear that the huge potential of mobile marketing has yet to be realized.

9.5 SUMMARY

We hope we have provided a few useful ideas and examples of effective launch strategies. The launch is often the most expensive part of a project, and it is when customers are first exposed to our creation. It is critical then that it goes well. We have shown that it is important to first decide what it is we want to achieve with the launch before we go headlong into commissioning expensive TV ads. We must be prepared to consider alternative business models and indirect revenue streams. Price will always be a barrier to use, but we can minimize this barrier by making sure people understand how much they will be paying. There are huge opportunities for collaboration between the mobile and media industries – we are only just starting to realize the opportunities. We have only touched on using the mobile channel for advertising and promotions, but this we believe has enormous potential. It may eventually dwarf today's revenues from ringtones and games.

10

Feedback and Quality

10.1 HOW ARE OUR SERVICES PERFORMING?

Do the customers like what we have launched? Are they not only using it but also recommending their friends and colleagues to try it? Are our partners in the value chain aware of what is working well and why? Looking at this area we find several very different views. On one hand you find media companies that build their entire business model and creation of new offerings around customer take-up (e.g. radio or TV ratings), and on the other you find mobile operators with massive amounts of network performance data. Also within the operator segment you find large differences between different companies, but also between the legs of the same organization (e.g. marketing vs operations). Additionally, the need for quality of service and customer feedback is related to the extent to which a company is focused on the success of its new services and retention of the key customers. As outlined in Chapter 2, there has not been much need to understand customers as long as voice and SMS have sold themselves. Similarly, if new services are not a core part of the business then the need for feedback on those is of course smaller (generic marketing vs one-to-one marketing).

This is an area without one true way to do things, and the solutions vary enormously, depending on the chosen approach. Consequently, we again focus heavily on case studies and real-world examples to aid understanding.

10.1.1 Keeping promises

One of the reasons we need to know how things are performing is to make sure that the customers get what they paid for. We have earlier seen the importance of taking away barriers of entry for new services and making them easy to use and easy to pay for. Now it is time to make sure that this is true not only during the definition and launch phase but also throughout the lifecycle. If some prepaid customers

Mobile Media and Applications – From Concept to Cash: Successful Service Creation and Launch   Christoffer Andersson,
Daniel Freeman, Ian James, Andy Johnston, Staffan Ljung
© 2006 John Wiley & Sons, Ltd

are experiencing difficulties in paying for their daily cartoon or their video services do not work for certain devices, you need to be alerted about it to be able to take action.

While this is key for all services, it is perhaps most important in the enterprise segment, where contracts are sometimes (and increasingly) tied to service level agreements (SLA) that outline the agreed performance and quality. As an example, a company using a wireless e-mail solution might have defined quality indicators so that 95 % of the time a GPRS/3G connection is attempted, the customer should be seamlessly connected and get an IP address assigned [packet data protocol (PDP) context activation]. This is increasingly important as the networks come equipped with more advanced QoS mechanisms, an evolution that is happening step-by-step. As mentioned in Chapter 4, there is a challenge in implementing cast iron guarantees of performance through QoS mechanisms, but at the same time it represents a strong opportunity for operators to differentiate their offerings. Consequently, in order to safeguard operator obligations to a customer, it is necessary to maintain a strong understanding of the true performance that such customers (for example enterprises) are expecting. We will return to examples of how to achieve this in practice below.

Finally, with many of the emerging services comprising components from various players, e.g. an application from a developer and content from a media company, it is important to keep track of usage to determine revenue shares and proper accounting. Operators are increasingly asked to report service performance to these external parties, but this accounting is often not satisfactory for the content providers. Actually, we have seen content providers who have stopped both operator and media partners from using subcontracted content for this reason – the accounting has been too bad. If the accounting from those providing the distribution of the content (operator, media) is insufficient, it is impossible for such a content provider to have a proper accounting relationship with such subcontracted providers that they normally use.

10.1.2   The importance of feedback

One of the most common mistakes in the world of mobile services has been developing offerings without proper customer feedback. The more we know about our customers the more natural it becomes to demand an understanding of how each segment is using the offerings. Just as no one wants to drive a car on an icy Swiss mountain road blindfolded, the whole mobile service value chain is starting to demand to see how things are – only using this feedback can proper decisions be made on what to do next.

One of the most significant aspects of Japan's early success was the introduction of feedback between the customers and the different parts of the value chain. New offerings were put onto the market based on the impact of the previous ones. In the rest of the world we have been so excited with the new technologies, or perhaps keen on catching up with the Japanese, that we have taken large steps at a time. The Japanese approach has been to take many small steps, improving the offerings gradually instead of spending big marketing dollars on bringing out the 'next-big-thing'. The Japanese word *Kaizen* (small improvements) exactly encapsulates this simple but efficient approach to enhancing an offering step-by-step, here illustrated by a Japanese aggregator:

"You know, people will not wake up on the morning when 3G is launched, yawning that they now need 3G applications. What people got from us that day was what they used and appreciated the day before but a little bit better. The fishing guide now includes higher quality videos of the top fishing spots and customers can send in their own videos or high quality photos of their catch. It's that simple, nothing dramatic."

The easy part is to understand that feedback is needed but then comes the question – *how* can it be done? Part of the answer lies in the question *who* is it for?

10.1.3   Who are the receivers?

A key aspect in determining what to use the feedback for is determining what the target audience is. In Figure 10.1 we have illustrated a number of possible receivers from the mobile operator perspective (we will look at other perspectives below, such as media). Traditionally the network view (performance management) has been dominant

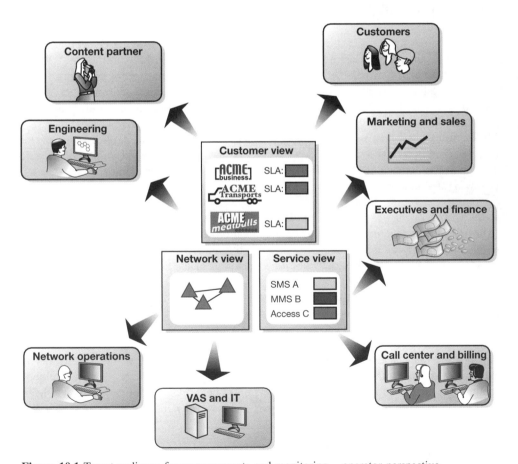

Figure 10.1 Target audience for measurements and monitoring – operator perspective.

and most of the log-files and measurements have been handled exclusively by the technical/engineering/operations departments.

As the value-added services become increasingly important, marketing and product management are growing to become equally or even more important receivers of such information. Knowing how things perform is one of the criteria for deciding the future of the service offering. Launching a new game depends on the performance of existing games and possibly other related offerings (e.g. Spiderman images and video clips if one is considering launching a Spiderman game). Additionally, finance needs to know whether service level agreements for customers and partners have been met when negotiating prices with enterprises and executives want to have a high-level view of what services are performing well and why. This is clearly in line with the convergence with media, where you can bet that the executives of a TV channel knows how their key programs are performing and who is watching. A TV show that does not get the viewer's approval is removed or improved, and commercial radio channels get higher advertising revenues when the listeners tune in.

10.1.4   State-of-the-art operator solutions

'The service has a 97 % success rate, the MMC (multimedia messaging center) node has only been restarted once this month, so the new patch seems to be working fine'. While this might be exactly what the operations and maintenance manager was asking for, it might not be the information that the marketing director or business unit manager (not to mention the CEO) needed in order to decide what features or content to promote next. This area, of measuring the technical performance for individual nodes (WAP gateway, MMC) in the network, is commonly called *network performance*. Aggregating several individual components to look at how customers perceive it is called *quality of service*. The standards bodies 3GPP[1] and International Telecommunications Union (ITU)[2] use the following definitions:

- *Network performance (component view)* – the ability of a network or network portion to provide the functions related to communications between customers
- *Quality of service (service view)* – the collective effect of service performance which determines the degree of satisfaction of a customer of the service

Often, network management is used to describe the management of the network performance (component view) and service assurance/management or customer management for quality of service (customer view or service View).

In the context of managing service performance, the unreliable nature of the wireless network often is raised as an issue. The related optimization of wireless networks is a recognized 'art' with various enterprises specializing in various tools and techniques for tuning and measuring performance of wireless networks. Historically this has been focused on network performance like analysis of low-level radio properties in the mobile

---

[1] 3GPP specification UMTS 22.25.
[2] ITU-T recommendation E.800.

network, drive testing (going around town in a van equipped with sophisticated measuring equipment) as well as extensive analysis of data logs from network nodes. For more detailed discussions around these areas, we recommend Gomez and Sánchez[3] and Andersson[4] as further reading. What we have found is that detailed aspects are necessary, but the big gains are found in taking an end-to-end approach.

When it comes to value-added services there are many new aspects that one must consider when evaluating the performance and accessibility of a service – especially since so many new nodes interact on the network side to complete a service. Consider the many extra nodes that are required to send and deliver an MMS message as compared with SMS. The interaction between different service layer nodes, elements in the radio network and the device creates a delicate balance that is only as good as the worst link in the chain. As media and content become part of the chain, we must also look at their contribution to the overall QoS. Part of this contribution comes from how these industries (TV, radio, etc.) work with feedback today, which we will look at below. However, since the mobile operators are the ones running the networks, we start by looking at this end – what are key issues around monitoring and assuring performance for an average operator?

Given the way these service networks have been created, adding one server/gateway after the other over time, it is natural that the systems for O&M look at the systems as separate entities, e.g. the WAP gateway is monitored as a separate system from the video gateway. This means that each node (system) is handled as a homogeneous entity including hardware, operating system, middleware and the application as such. The only integration is towards the O&M terminal for different systems (via a graphical user interface). This makes it possible to reach all systems from an operations terminal, but it is not a simple task. Each system might have additional tools for trouble-shooting purpose within that system only. As an example, a top-tier operator even in 2005 manually had to compile reports from a large number of node-related performance reports in order to get a high-level view of how the service network was working.

The device and customer experience side is seldom included in the overall chain of measurements. In this sense, the horizontal approach is missing and end-to-end performance for the services is usually not monitored. Consequently it is not easy to fix those faults that pass unseen when only looking at one or two nodes or subsystems.

Even in those cases where end-to-end service assurance is measured, the links to the business processes is usually lacking (service continuity, root-cause analysis, etc.). The work is labor-intensive, costly and difficult to expand as the network grows. Organization-wise, those creating measurements are often those who previously implemented the systems and therefore to some extent caused the problems. Sometimes a quality or customer relationship management (CRM) department exists (note that this alone is not the solution), but its level of impact, pro-activity and integration in feedback processes varies.

10.1.5 Understand the network and start simple

Without looking at all aspects of quality management and product lifecycle management, we would like to stress some important input to measuring QoS. The mobile network

[3] G. Gomez and R. Sánchez, 'End-to-end quality of service over cellular networks'. John Wiley & Sons, 2005.
[4] C. Andersson, 'GPRS and 3G wireless applications'. John Wiley & Sons, 2001.

is a shared resource and, even though advanced mechanisms insure the quality for our phone calls and services, there is always a limit to its capacity. This means that customers can have trouble making calls or sending e-mails if the network is overloaded, like at a football game. While most people are used to these effects, they can be devastating if an operator has an SLA with an enterprise that uses the network for mission-critical business applications. As an example, a shipping company used the mobile network for logistics applications but always had big problems with the reliability in the mornings. An investigation showed that the local radio stations tended to have lots of competitions (with customers sending in responses via SMS) in the mornings, which resulted in an enormous SMS load. Since SMS uses the control channels of the network, this traffic not only lowered SMS performance but also the ability to make voicecalls (as the control channels are needed for call setup). While the solution here was to dedicate an SMS center to this particular customer, more interesting conclusions can be made about the need for good network understanding. What do we need to know about the network before launching new services and introducing solutions to manage the quality?

We see two main areas to consider:

- *Network model (component view)* – these models are rather static, and show how the network is built and where its strengths and weaknesses are. Maybe the transmission capacity to a remote province is slightly low and some other area has certain weaknesses in the radio coverage.
- *Usage and traffic model (service view)* – these are more dynamic and illustrate the nature of the traffic in the different parts of the network. What are the peak hours and how large is the traffic load? An example is how some business districts have enormous traffic load during working hours but very little at night-time. Another example is the so-called 'potato-peak', when workers call to say they are on their way home (named after the fact that some customers claimed they did this to ask the spouse to start boiling the potatoes, which does not sound like a celebration of the equality of the sexes).

Clearly it makes a difference if an operator has created links between these models, which are often owned by the chief technical officer (CTO), and those planning and launching services. As an example, a marketing campaign for a new service could not be focused mainly on an area where the capacity is temporarily limited. As we saw in Chapter 1, new business models require new processes and commitment to executing them across the organization. The service offering needs to be connected to these business processes and this is even more evident when talking about service level agreements with enterprise customers – you have to know what you have to offer. This was shown when a large logistics company approached a large and well-developed operator, asking for an agreement of the quality for its field workers around the world. The operator was of course thrilled and very interested in signing such an agreement (for big, big money), promising that it could assure the radio quality. However, it turned out that this logistics company could easily prove that it had a much better view of the radio quality than the operator (based on measurements of their fieldworkers' communications), who therefore would never be able to live up to such agreement. There are solutions for such requests (we will look at a similar case later in this chapter), but they require the homework of both network/service models and usage/traffic models.

Some of these examples (like the global logistics company above) are rather extreme and, as with many things, you have to learn to walk before you try to run. We have seen the best results when starting small, by looking at a specific service (e.g. MMS, SMS), which is important to the stakeholders (e.g. customers and/or management) and generates payback. If this first step is done with a good strategy and well-planned architecture in mind, the next step can be built on the successful result and feedback generated. In line with this thinking, the preparations (like the network model) should also be done per service to make it manageable.

10.2   END-TO-END SERVICE ASSURANCE

Getting into the *how* and outlining some example solutions, the key is to keep the benefits of monitoring the technical systems (network performance) but additionally introduce means of obtaining the big picture on how things work from the customer point of view (quality of service). We will use the term *end-to-end service assurance* when looking at the quality of service from a service view (Figure 10.2).

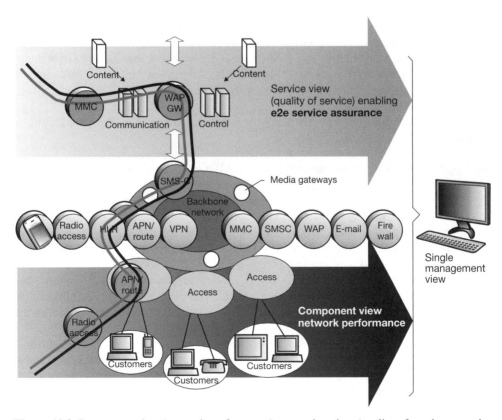

Figure 10.2 Component view (network performance) vs service view (quality of service – service assurance).

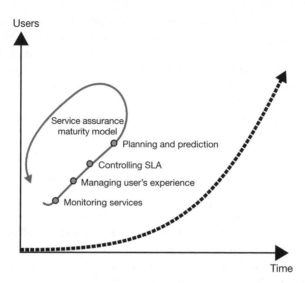

Figure 10.3 Service assurance maturity model.

End-to-end service assurance takes the high-level view of individual services and combines the input from many network-level components. Usually one starts with implementing monitoring capabilities and then increasingly manages the customer experience until it can be guaranteed in SLAs. This evolution or maturity model is illustrated in Figure 10.3. From there one can go back to the first step to improve the solution and extend the model to new services.

Putting this into practical terms, here are the normal steps are taken when implementing a service assurance solution all the way:

(1) Decide *why* – this sounds obvious but it is still a common mistake to omit this step. Is it to be able to catch alarms, faults or other indications that things do not work? Perhaps you want to understand what customers think of the services right now? Or maybe the main target is to continuously (in real-time) receive updates on key indicators of performance? With some of these questions it will be obvious whether qualitative or quantitative analysis is needed and this narrows down the scope as well as defining who will receive the information.

(2) Decide *what* you want to look at – Is an overview for the whole portfolio needed or can you start with one or a few key services? The most common requirements we have seen here are on one hand to know how selected services perform and on the other hand to be able to track individual customers.

(3) Define key quality indicators (KQIs) for those services, from an end-customer view. For MMS this could mean the time it takes from accepting a notification of a new incoming message until it can be played in the device. For e-mail access, on the other hand service accessibility (including coverage and capacity) could be the KQIs.

(4) A service model is developed that indicates dependencies between the component and service view. This step is the link between the higher lever needs and concrete ways to implement then (part of the homework we discussed in the previous section).

(5) In order to determine what measurements are needed, the reports and alerts for monitoring KPIs are defined. For enterprise services, it is common to connect the KPIs to the SLA that has been signed between the service provider (mostly an operator) and the enterprise. The monitoring system can then be used to illustrate how this SLA is fulfilled. The mapping from KQIs to KPIs can in some cases be quite complex and in this book is only described from a high level.

(6) Finally, the measurement methods are decided and implemented, realizing the needs outlined in the steps above.

Let us now assume that we have now decided why we are measuring things and the next step (number 3 above) is then to look at what the experience is that we want to assure for our customers.

10.2.1 Measuring the customer view – defining the KQIs

Defining a quality measurement for a service is specific to each and every service and is naturally open to debate. In any case, however, it is paramount to define a measure that relates to the critical aspects of the quality that make the biggest impact on a customer. What will make them feel satisfied with the service and what would make them complain? It is no good measuring simply the success of a service – a WAP page may load successfully after 30 s of loading; however, from a quality perspective that service is a failure. Defining the most important use-cases and scenarios is one way to narrow down the scope to the aspects that are most important to the customers.

Looking back to earlier chapters we can see many of the key aspects of service design that relate strongly here when measuring quality – here are some examples:

- E-mail and corporate access (enterprise perspective) – frequently the network accessibility is the key KQI for a company that gives its employees laptops with GPRS/3G connectivity. The employee must be able to rely on and access the network when needed. A consideration here is that in mobile networks it takes more time the first time a connection is set-up (PDP context activation to get an IP address, etc.).
- MMS – many operators are satisfied with the amount of time it takes for a MMS notification to arrive, but if the message is not fully downloaded, is this a true measurement of service quality? The time it takes from sending a message to the recipient being able to fully view and 'experience' the message is a key indicator here, as we will illustrate in a case study below.
- Video telephony – diverging from traditional voice services, the quality indicator here is not the time needed to connect a call, but the time it takes for the video content to appear on each participant's display at the start of the call, depending on network performance and the device's video rendering speed.
- Streaming services – for video streaming services, the frame rate is key, which is rather tricky as it depends on the video coding format. As interruptions or temporary decreases in bandwidth can have a large impact on the experience, these are important quality indicators. Here the buffering in the device can sometimes compensate for variations, which can be a challenge to include in the solution.

As mentioned earlier, the definition of KQIs can be quite complex and as always it is worthwhile to start simple by collecting input and feedback from the stakeholders, whether it is management, customers or someone else.

10.2.2 Defining a service model

In order to map these KQIs into concrete measurements and monitoring, you have to map the chosen service into network elements. The service models show how services impact network elements and vice versa, like how SMS impacts the control channels in the example above.

The setup of the service, how it is used and what parts it affects are key elements to illustrate. Continuing with the mobile corporate access example (employees with laptops, PDAs or smart phones that can access the company intranet), we saw how the network accessibility was key. Looking at Figure 10.4 we see how this is realized in a service model which is primarily dependent on the radio and core network performance.[5] The

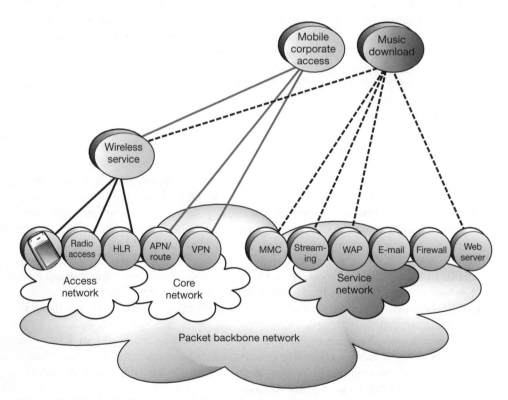

Figure 10.4 Example of service models – corporate access and music download.

---

[5] The mail server and its surrounding are of course key, but often that is managed by the enterprise itself. In this example the operator SLA did therefore not contain the mail server performance.

APN seen in the figure is the link between the operator network and the enterprise while the VPN creates a communications environment specifically connected to that company (short numbers, etc.).

Figure 10.4 also shows a service model for a music offering, which allows customers to be updated with news articles regarding pop-groups, view video samples from their latest music videos and download to samples as well as full songs. This service is much more dependent on the service network components, like WAP gateway and web and streaming servers. For this kind of service, we will increasingly see demand for connections to other partners in the value chain, like content providers.

10.2.3   Key application KPIs in wireless networks

The challenge is now to use the service model to find appropriate parameters, KPIs, for monitoring the quality. This means that the things that customers perceive as quality (KQI) are translated into very concrete measurements (KPI).

Since the measurement methodologies and solutions vary greatly depending on the existing solutions of that particular operator and the needs, we will describe this mainly using case studies. The first case is for MMS, used to illustrate a typical approach for consumer services.

Case study: E2E MMS performance monitoring – implementation
Let us look an operator in Europe, which introduced MMS-based services and saw the need to monitor MMS operational service quality. Key questions included:

- What is the usage of the service?
- How many messages are sent mobile-to-mobile/mobile-to-e-mail?
- What is the main reason for failures when messages are not delivered?
- What is the average delay (QoS) in delivering messages?

The first step was to implement high-level monitoring functions, like a dashboard where one could quickly get an overview of the key aspects and performance reports.

For monitoring and assurance it is important to differentiate between the indicators that customers perceive the most and those that are more for the operator to understand the overall performance. In the MMS case, we concluded that 'the time it takes to send an MMS' and 'how long does it take to download a message' are appropriate measurements.

With the enormous number of measurements possible, the challenge is picking the right things to look at. As an example, for every MMS sent, 10 different charging data records (CDR) are created. Looking at the right things makes life a lot easier. A sample of the KPIs selected for this particular example is shown in Table 10.1.

The KPIs are refined and processed to make sense from a service management view and the service level manager reports real-time monitoring of service level objectives to the desired customer management systems. This is shown from an architecture point of view in Figure 10.5.

Table 10.1 Example KPIs for E2E MMS service assurance

Synthetic transaction monitors	Performance agents (at MMS node)	Fault management agents (at MMC node)
Response time	MMS download delay, send delay (s)	CPU load
Availability of service	MMS download throughput, send throughput (bps)	Memory utilization
Response time	No response from WAP gateway (%)	Process status information
Availability of service	Mobile terminated/mobile originated failure ratio (%)	

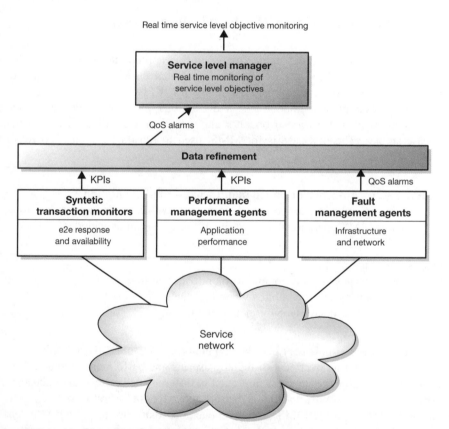

Figure 10.5 Architecture view of E2E MMS service assurance.

There are many reports and customer management tools that can be used to format and visualize the output information. In Figure 10.6, we can see one example that proved valuable in this case, namely the distribution of the messages per content type (simplified version showed).

For another example operator the use of an MMS E2E service assurance not only led to improved monitoring and assurance but also direct improvements in quality, with

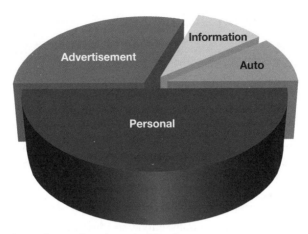

Figure 10.6 Example service performance visualization.

€1.8 million of annual cost savings. In that case, the improvements identified were mainly in the areas of provisioning failures, network performance issues and a fault in the MMC that led to 'message-not-found' errors.

Implementing solutions for monitoring of other services does not differ dramatically from the example above, but the choice of KPIs does. The impact on customer perception of quality varies greatly for different application types, which is illustrated in Table 10.2. This table shows how, for multiplayer gaming, throughput and network speed are less important than for streaming services.

In earlier chapters we saw how latency (the time for a message to loop forward and back between two entities, i.e. a mobile device and application server) is key for multiplayer games. As a matter of fact, this has such importance that the players are often (for fun) categorized according to their respective latency (commonly called ping-time). Above 100 ms they are called 'HPW' or 'high-ping-whiners', just because these players tend to complain a lot about their high-latency disadvantage. The players with latency under 50 ms are called 'LPB' or 'Low-ping-bastards'. Clearly, latency must not be forgotten when looking at quality of service for multiplayer gaming applications.

Table 10.2 Key aspects for different mobile applications[6]

Service	Latency	Session stability	Throughput	Predictability
WAP browsing	High	High	Low	High
MMS	Low	High	Low	Low
Streaming	Low	High	High	Low
Gaming	High	High	Low	High

---

[6] Source: Vodafone at HSDPA Conference 2005 and Ericsson.

10.2.4   How to measure and monitor

Monitoring of services can be made in two ways – passive and active. Passive monitoring relates more closely to traditional means; however, instead of analyzing network node logs, customer traffic is captured unobtrusively and data is collated to establish an understanding of the performance of overall sessions. In this way it is possible to construct an impression of overall service performance. This approach can provide excellent high-level data mining capabilities to expose the most popular services but also to understand the highest causes of service failure on a device-by-device basis – often due to customer mistakes such as entering incorrect phone numbers or particular phones not being correctly configured.

Active monitoring relates to the use of test probes (often distributed geographically across a network) that actively send data as if they were genuine customers. These probes are often best used to control genuine mobile devices that imitate the behavior of customers and execute complete service transactions, in which case the customer's perceived performance can be instantly derived and reported on. These approaches are complementary to each other and to traditional node monitoring. When used together they can provide a service provider with a total overview of what is going on.

10.2.5   Including radio and core networks

We have looked at corporate access as a driver for measuring the radio and core networks, and now we will dive into its implementation aspects. For example, ensuring the reliability of the mail server does not help much if the customer in question is constantly experiencing an overloaded radio network in the area that he or she is working in. The case study below describes how this challenge can be addressed.

Case study: insuring intranet and e-mail access quality for key corporate customers
Corporate access to e-mail can be implemented in several ways, for example via dedicated mail gateways (e.g. Blackberry). This means that a dedicated solution for the wireless access is implemented that is tailor-made to work well with the wireless environment. In this case study, another approach was chosen, based mostly on the network level, by ensuring that the whole intranet access was enabled also for wireless customers (VPN-style). Rather than focusing on e-mail alone, this also enabled web access and most other internal applications. With very explicit demands to ensure that the service quality measured included radio network aspects, the KPIs in Table 10.3 were identified.

With these measurements it was possible to meet the need to monitor the service from a network point of view. The wireless access measurements via probes enabled tracking of the availability of the network while the backbone systems delivered data around the delays and lost data packets (Figure 10.7).

The radio part is mostly the trickiest, as external probes cannot measure all protocol levels in the communication without interfering with the traffic. This is delicate, as you probably do not want to introduce monitoring mechanisms that lower the overall performance. Real-time performance monitoring (RPMO) is a solution to this, where the radio network controller (RNC) and operational support system are updated to look at the traffic without interference. While this requires a change in the access network (the RNC),

Table 10.3 Example KPIs for a corporate access solution

From radio network operation support system	From nonintrusive probes	From packet backbone operation support system
Operation availability	Radio access network (RAN) availability (business district)	IP VPN packet loss
		IP VPN round trip delay IP VPN availability Threshold alarm reports
User database and radio base station uptimes Threshold alarm reports	PDP context activation success time	IP VPN round trip delay
		IP VPN availability Threshold alarm reports

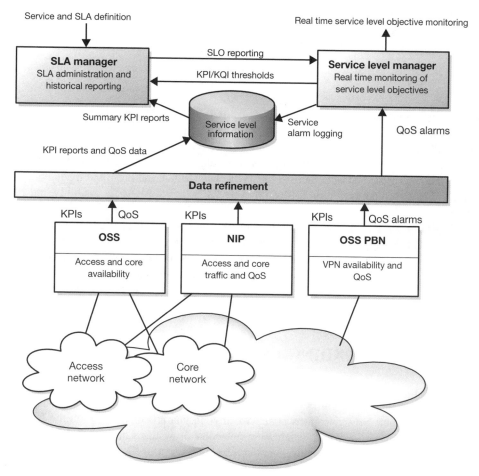

Figure 10.7 Architecture view of corporate access service monitoring.

SLA Monthly Report

ACME Meatballs corporation
Santa Valley 4
North Pole, NP 28006

July 2005

Service name	Threshold name	Value	Units	Actual value	SLA
GPRS/3G corporate access	RAN availability (business district)	95.00	%	97.00	Met
	PDP context activation success	95.00	%	89.00	Violation
	PDP context activation time	2	s	2.8	Violation
	VPN-223 packet loss	0.5	%	0.3	Met
	VPN-223 round trip delay	95.00	%	98.00	Met
	Operational availability	99.50	%	99.55	Met
	Operational mean time to repair	30	min	18	Met
	TRB maximum number of outages	5	outages	1	Met
	TRB maximum outage length	30	min	18	Met
MMS	Response time mobile to mobile	2	min	0.6	Met
	Response time mobile to e-mail	5	min	1.5	Met
	Response time e-mail to mobile	5	min	1.7	Met
	Operational availability	99.00	%	89.9	Violation
	Operational mean time to repair	60	min	46	Met
	TRB maximum number of outages	5	outages	4	Met
	TRB maximum outage length	60	min	30	Met

Figure 10.8 Illustrative SLA showing mobile operator reporting to an enterprise customer.

it also increases the performance greatly. An alarm of traffic overload that is normally generated after 45 min instead comes after 2 min – a huge difference if you want to use this feedback to take action.

Figure 10.8 shows how the KPIs for this case study can be summarized in SLA reports, where the enterprise customer can see clearly to what extent the promises made are being kept. The network accessibility is illustrated by KPIs like the RAN availability and PDP context activation success.

A business unit for enterprise customers or the chief finance office (CFO) organizations often manage this type of SLA, while the technical implementation is done by technical organizations (and increasingly CRM). This returns us to the fact that we have emphasized several times – success comes from forceful implementation of the technology in close connection with adapting the organization towards new, cross-functional, business processes.

10.3   THE EXPANDING END-TO-END

Given that many services now (and even more in the future) run partly over the Internet, where video content or a game server could be located, this part of the chain has to be taken into account. Now questions arise: can anyone guarantee anything on the Internet? How can I assure performance when I am not controlling the whole chain? Simply put, this usually means that one player cannot control every part of the service delivery chain. However, the service assurance approach described above is still valid, but sometimes complemented by active monitoring that makes it possible to check the customer experience view without knowing all the components along the way. With this approach, there are also solutions to assuring the network performance, e.g. by getting a leased line (dedicated to the service) between an operator and an enterprise customer or partner.

10.3.1   Active performance monitoring from the customer perspective

As described above, active monitoring differentiates from passive monitoring in that customer usage is emulated to stimulate the systems. This way you can automatically test how things would behave without knowing much about the servers and networks that are passed along the way. In the case below, we will look at a large service provider (that is not a mobile network operator) and its way of monitoring performance.

Case study: performance monitoring from the customer perspective with a global service provider
This particular service provider has a global footprint and allows customers to download ringtones, videos, games and other services via an advanced WAP portal. The service was very appreciated by customers but, whenever someone complained about broken links or slow download, it was hard to know what the cause was and what actions were needed. Therefore, we jointly looked at different methodologies to get a better view of what was working from the customer point of view and what needed improvement.

In this case, active monitoring was used in diverse geographical locations to isolate performance problems relating to services in that specific country.

Initially, the portal technology seemed to be functioning properly and only after repeated active tests were analyzed were serious flaws discovered in the flow of the application. These became obvious, and additionally numerous incorrectly defined links to content items were found. Now it is possible to do direct analysis of specific areas of a portal, having complete control over the use of a specific mobile device's profile and improve the detection of faults. Before this project, the global portal stability was unknown (it proved to be 68 %) as well as the availability of content (average 80 % accessible). After the project, these figures were significantly improved, with over 90 % portal stability and close to 100 % content availability.

Here we see how active measurements improved the performance and a customer-centered monitoring approach showed the issues quickly.

10.3.2   Measuring customer perception (qualitative)

No matter how good our solutions for measuring, how much each customer segment is using services and other quantitative data, there are still things that will remain unknown. Why did that customer stop using the multiplayer games after the first month? What content is the high-usage customers missing in the video service package? Why is the usage so low for the service that our main competitor is generating the major part of its data revenues from?

For this, qualitative measurements are a help. We will look at two different methodologies: customer care/helpdesk and user groups/surveys. By 'customer care' we refer to the call centers and helpdesks where customers can phone in (or visit a store) to ask questions or complain about the services (customer care is sometimes also used to describe the operation and maintenance systems or organizations). The customer care function is usually a reactive entity, which captures the issue when the customers have already become frustrated. Our experiences with various mobile operators show that, when customers have some issue of relevance, they only call customer care around 15 % of the time. This means that most of the issues are either solved by the IT guru friend/family member or not at all (the customer gives up). Equally important, the feedback from those frustrated 85 % of the customers with problems remains unknown. Another issue with the customer care units is the cost associated with handling these calls from the customers. Some €5 of operational cost per call to customer care is a common approximation.

To a large extent, customer care personnel do not have the tools needed to discover what problems the customers are experiencing, which makes it hard to do anything but to remain re-active – waiting for the next call with complaints and problems. One solution to this can be strong customer care groups that have a variety of phones on hand to try out the services that a customer is having trouble with to compare experiences. Other examples include creating a connection to the service assurance output and the usage and traffic models used. This obliges organizations to train support staff more fully on the use of various services. The direct customer interaction that customer care units have is a valuable asset that is tricky to manage in an optimal way.

Actions to lower customer care costs and free up these resources to deal with issues that add more value to the operator/service provider as well as the customers include:

- making sure that single sign-on is implemented to avoid the large number of calls around lost passwords – no log-in should ever be needed;
- ensuring that automatic device configuration is implemented to always detect which devices are un-configured or wrongly configured and automatically send the right settings;
- connecting customer care to the feedback loops created, e.g. service assurance, and making it possible for them to give input to the service offering evolution.

Overall, the customer relationship management functions will play an increasingly important role in the move towards creating one-to-one marketing and new business models.

User surveys or user-groups represent another way to measure qualitative customer feedback by interviewing the customers directly. These can comprise anything from the 15 neighbors that the small developer gathers to obtain feedback to a new service to the thousands of interviews that some of the big vendors and operators perform annually. To make a generalization, the most common issue with this kind of measurement/survey is when it is not done at all. Whenever selected customers explain how they use services and what they feel about them, you can capture much valuable feedback. The main risk is selecting too few customers that do not represent the target group for the offerings in mind. If you plan to offer a casual gaming portfolio (Tetris, Solitaire, etc.) and try it on a bunch of hard-core gamers, you are likely to draw the wrong conclusions ('It would be great if you could play with your mobile against your friend on his XBOX 360', while the typical target customer was an 18–35-year old woman who had never owned a game console).

10.3.3   Media industry – understanding usage is everything

As we saw in Chapter 1, the commercial radio stations are dependent on the revenue from advertisers and therefore of course have to insure that they can prove that people (and the right people, advertiser target groups) are listening. The same goes for advertising-based TV channels, which we will take a brief look at below. As mobile services become more and more media and applications through a new channel, the media industry's knowledge of how to generate feedback is growing in importance.

TV ratings is a topic that stirs up emotions among most players in that industry. Usually done by independent research institutes, these are critical to the business but there are also constant debates about their quality, precision and pricing (several of these independent players are in close-to-monopoly positions). The research institutes are different in different markets, but their approach and methodologies are similar. We will therefore look at the US market and Nielsen Media Research (Nielsen ratings) as a means of measuring viewing patterns and TV ratings.

To start with, there is no simple way to track what you and I are watching on TV. Unlike a web site or magazine, you cannot go into a log-file or the sales statistics to see how many viewers there were for a certain channel or show. Therefore, Nielsen and other research firms estimate the audience via sample audiences and statistics. The concept is commonly called 'TV-ratings', but this does not indicate a qualitative or subjective view of a program but rather just how many watched it.

In the USA, Nielsen typically uses samples of 5000 households (around 13 000 people) and, despite this only being 0.005 % of the 99 million US households, it still gives a good view of the audience turnout (which those of you who still remember some of those statistics you once learned at school, might see). These households are selected randomly and equipment is installed in their homes (connected to TV, VCR, cable and satellite boxes) in order to measure what was viewed and for how long.[7]

As we have seen, the household viewing figures are key for knowing how appreciated a program is and whether to change or remove it. For advertisers, on the other hand, the demographics are key, which are handled by boxes where the family members indicate when they are watching.

What can the mobile channel add to media companies' ability to follow usage and get feedback? A lot, it turns out. We have seen throughout the book how the mobile channel is interactive by nature and how there exists a very direct relationship to the customers (and not only the households). SMS voting and payment (e.g. donations) has become a standard component of television in many countries and we are starting to see the subsequent steps being developed. The interaction capabilities of the mobile channel make it excellent for facilitating feedback and continuous improvement. It is rather easy for an operator to look at who is sending SMS at what times, which then can be used to draw conclusions about who is using what services and when. The proportions of such samples from the mobile channel are often quite different than the 5000 TV households mentioned above. As an example, Cingular Wireless noted 41.5 million SMS during its 12-week voting period of "American Idol" in 2005,[8] which should do for a decent sample.

---

[7] Source: www.nielsenmedia.com.
[8] Source: www.slashphone.org, 27 May-2005.

Even higher potential will come from combining the feedback from different channels, in this case mobile and broadcast TV, both quantitative and qualitative.

A common feedback that we have found in cross-media offerings like mobile TV is over-promising of new services. Customers are familiar with the concept of watching TV and this normally means hi-fi sound, smooth movements and a huge screen. If the mobile offering is positioned this way, it clearly has a negative impact on customer perception.

Case study: qualitative feedback from Italian mobile TV customers
A qualitative study carried out in 2005[9] of the Italian mobile TV experience used in-depth interviews with early adopters of two competing service offerings in that market. Example of this feedback include:

- The promotion should be adjusted so as not to over-promise the experience. Customers of one of the operators were happier than the rest just because that operator had transmitted realistic expectations.
- Sound quality was an important factor that could be improved. This was initially not considered an important factor.
- Customers found the waiting time for switching channels to be key. Consequently, this is a suitable KQI and KPI for such services (should not exceed 3 s).
- A successful first trial is a vital success factor. Demo stations should have been more extensively deployed across the distribution network.
- Several customers wanted to access the service in situations where there was poor coverage. This gives concrete insight to address coverage problems and understand where the coverage has the most value to the customer.

This kind of feedback can then be used both do decide how to evolve and improve the offering (for TV and mobile) as well as defining which parameters to measure in a service assurance solution (e.g. channel switching time).

10.3.4   Developers – evolving the fastest

During the infancy of mobile services (which we are still in), the small and medium-sized developers have been driving much of the development and the innovation. This also includes how customer behavior and feedback are traced and used. The tough times developers experience in trying to work with the big players forces them to fight hard to differentiate and find innovative ways to add value. Interestingly, many entrepreneurs we have worked with have additionally worked in consumer marketing or retailing, which has been shown to give a good background for success in this area.

With the ringtone business being one of the early success areas, we have chosen a case study from this area to illustrate this.

Case study: MobileHits mobile music content and customer behavior adaption
MobileHits (now owned by Aspiro) was one of the pioneers in the European ringtone market and acted as an aggregator between the record companies and the operators. In

---

[9] Per Löfqvist and Jonas Agartz (Ericsson Consumer Laboratory and Argur Market Research), 'The Italian mobile TV market', 2005.

addition to the basic aggregator offerings (content reformatting, brokering, etc.), Mobile-Hits designed a platform that took large steps towards understanding customer behavior and made it possible to make accurate recommendations for customers based on previous downloads. This meant that all usage could be monitored in real-time and new campaigns and offerings were always based on this instant feedback. As an example, a promotion was made to customers between 18 and 25 who lived in the southern most county of Sweden (Skåne) and who had a preference for rap music ringtones. Those customers received an SMS saying that they would get a free T-shirt with a popular rap band if they went to a certain music store downtown and bought the new CD of the band in question. This SMS was sent out on Monday based on the MobileHits database query and by Wednesday, when the record company launched the associated campaign on TV the T-shirts had already sold out.

From the MobileHits case we can also learn that this kind of developer and corresponding media partners have a need to access more information about customer behavior and would like to collaborate with mobile operators to obtain more customer data. This includes whether a subscriber has a phone capable of video or high-quality music, for example. As the business and value chains evolve, we are likely to see more collaboration between the different players.

10.3.5   Offering more to enterprise customers

Can the measurements and monitoring solutions discussed in this chapter be used to increase the value even more for partners and enterprise customers? The answer is yes, and this is one of the areas we are likely to see growth in over the coming years. Take the example of all the content-to-person SMS that companies (content providers and enterprise) send out to customers. Today the ones sending out these SMS have their own database of phone numbers, uncorrelated with that of the operators in question. Consequently, lots of SMS are sent to phone numbers that are not currently in use and increase the load on the network. An operator with 10 million subscribers and 8 million content-to-person SMS per day had over 150 000 of those SMS not reaching the destination. The case study below shows an example of a solution to this problem.

Case study: bank and operator tighten collaboration
The bank in question had very high demands on its IT services to its customers and wanted to improve the capacity for its SMS notifications (customers get alerts when the debit card is used, for example). After some discussions it was evident that major improvements would be possible if a dedicated SMS center were allocated to the bank and its customers. However, even more improvements could be achieved by creating a connection between the operator customer database and that of the bank. Today, the SMS notifications to the bank's customers achieve a 99 % success ratio and arrive within 2–3 s of when the transaction is made. Many customers are actually surprised when they withdraw money from a cash machine and after only 2 s receive confirmation.

This is just the tip of the iceberg of what this kind of exchange of data could bring and operators can offer quality reports and more advanced connections to all sorts of enterprises and content providers.

10.4   KAIZEN – CONTINUOUS IMPROVEMENTS

In earlier sections we hinted at how the Japanese approach has contained something fundamental in its way of looking at service lifecycles that is often missing in other countries. The word *Kaizen* is sometimes used to describe part of this approach, but what does it really mean and what can others learn from it? Is it the way all services should be managed?

To understand Kaizen, we need to understand more about the Japanese industry structure. During the 1950s and 1960s Japanese industry went through major changes (evolution and via stimulation by the state) and it became very focused on consumer product offerings, like cars (Toyota, Honda, etc.) and electronics (Sony, Sharp, Hitachi, etc.). These companies succeeded in the highly unpredictable consumer market by investing heavily in product innovation and then testing on the market. A consumer electronics company like these can have three parallel development tracks for a new game console with internal competition on which to launch. Sometimes the winner is selected before launch and sometimes two to three products can be put on the market at the same time to let the consumers show which one is the best. The ones that are less successful are most of the time quickly discarded from the product line, while the winner is used as the basis for evolution into the next generation of products.

Case study: from DoPa to iMode

As an example, DoCoMo in 1997 launched the DoPa offering before launching iMode. It could give customers speeds of 28.8 kbps and some basic services. It seemed, however, that customers could not have cared less and DoPa was a failure. If Kaizen is about small improvements to what you already have, then DoCoMo surely must have tried to improve DoPa to make it successful? The answer is no. One of the keys to this philosophy is that you quickly scrap what is not working and improve what is working. Consequently, Mari Matsunaga was brought in to add a fresh, new approach and we have seen how she succeeded. Actually, the resulting iMode service offered *lower* speeds than the DoPa offerings, just because the team had set the goal to make the service simple and cheap. Thus, 9.6 kbps was enough.

From the initial success with iMode, the things that worked well were evolved step-by-step into better and better offerings. This not only refers to the services but also to the devices. A device in Japan normally has a lifecycle on the market of about one year. However, each device that DoCoMo offers comes in two versions, e.g. 903i and 903is, where the latter is a slightly improved version released 6 months after the first one. The sales of these then show how the new features in the improved phone (903is) are received and influence which features to add to the next generation device.

So how is customer feedback measured in iMode? While of course DoCoMo has its own, internal, ways, the most striking methodology for feedback generation across the value chain is the iMode menu. The menu of the portal is arranged so that the top-performing services are put at the top and then lower-performing ones towards the bottom. This means that the developers and content partners can look at the menu to see how their offerings are received as well as which kinds of competitor offerings work well. Additionally, the consumers can clearly see what the most popular services are, which stimulates them to try them. This has a similar effect to the Top-40 or Top-20 lists that the record industry is built around – people have been proved to be interested in trying things that others like. Communities are key.

What else can we learn from this case study and Kaizen in general? Clearly, the philosophy of improving services in small steps rather than dramatically fits well with the consumer understanding we have shown here. Changing customer behavior takes time and we need to move slowly, building on what people like and are used to. An interesting aspect in the move from DoPa to iMode is that DoCoMo used an approach very similar to the one we have shown for media companies to use (in Chapter 3) – select an excellent producer and give her or him the budget and freedom to create something that hopefully consumers will like. Actually, several successful consumer electronics companies in Japan have built their success on trying things out (and spending the money on that), seeing that as more important than doing extensive market research.

Summarizing some of the knowledge gained from Kaizen:

- New services should be small improvements on what already exists (that customers like).
- Close down offerings that do not work – trial and error.
- Improving an offering is not always about bigger, better and faster things, but improving the things that matter to customers. Here it was making it simple and cheap rather than faster.

As we are starting to see success cases of mobile services in the USA and Europe, we are also starting to recognize the use of Kaizen outside of Japan. A Japanese would, however, advise you not to try to create a huge methodology around this (as Westerners tend to complicate things), but rather just take it as a simple philosophy – simple enough to facilitate creation of services that are simple and attractive to use.

10.5   SUMMARY

Since you have created your mobile media and applications based on customer needs, it is natural to follow this up with good feedback mechanisms to find out what the customers are experiencing. Quality of service is a wide and sometimes complex topic where you need to start simple and evolve gradually. Service assurance enables you to monitor usage and agreements with customers in real time, giving a better overview of the quality and feedback. We can also expand this feedback towards the other players in the value chain in order for all to contribute to improving the offerings. In the end, this can create the continuous, small improvements that are needed to satisfy customers and grow the business.

Appendix 1

Take Five Consumer Segments

A1.1   YOUNG PIONEERS (P1)

Demographic   15–24 years old; slightly more males.

More likely to be single and still live with their parents.

Many are full-time students.

If working, they are more likely to have white-collar jobs or work in shops.

Personality   The world of the Pioneer is a world of action. Pioneers lead a multitasking existence – life is too short to do only one thing at a time.

They are in line with the latest trends in society – in fashion, technology, and new ideas – often without knowing it themselves. Pioneers are devoted to pleasure and seek challenges; they feel at ease in creative and dynamic settings. Pioneers are brave and daring – they like life to be varied. Owing to their strong sense of individualism, they want flexibility in life, in relationships and at work.

Young Pioneers are strong, open-minded, individualists who ambitiously explore life's potential. They like excitement and adventure and are more fun-oriented than Adult Pioneers.

Lifestyle   Young Pioneers lead a busy, outgoing life with lots of socializing with friends.

They listen to music – especially new and avant-garde – and like to watch movies, both in the cinema and at home.

Mobile Media and Applications – From Concept to Cash: Successful Service Creation and Launch   Christoffer Andersson, Daniel Freeman, Ian James, Andy Johnston, Staffan Ljung
© 2006 John Wiley & Sons, Ltd

They like to exercise and often participate in extreme sports such as bungee jumping and paragliding.

They browse in stores for fun, and tend to buy trendy clothes.

Young Pioneers are more likely to play PC/video-games and read cartoons than other segments.

They also attend evening classes, spend more time reading books than others, and are more likely to read horoscopes.

They choose to watch TV programs such as MTV, but also watch movies, comedies, soap operas and reality TV.

Technology use   Having up-to-date products is very important to Young Pioneers. They are interested in new technology because it is new and exciting, but also because it is fun and entertaining. They use new technology for entertainment, such as computer games and downloading music from the Internet.

They are heavy users of the Internet, especially e-mail, chat, ICQ[1] and SMS. They browse the Internet for entertainment and leisure information and information on products and about sports. They commonly download audio and video content and send jokes to their friends.

A1.2   ADULT PIONEERS (P2)

Demographic   25–49 years old, although most are aged between 25 and 39; slightly more males.

Well educated – better educated than most other segments.

High household incomes.

Work full-time as managers, consultants or in IT; many are self-employed.

Personality   Creativity and curiosity are important to the Adult Pioneer, as is simplicity – they like things that make them efficient and that are easy to understand.

They are slightly less fun-oriented and more rational than Young Pioneers and tend to be more interested in new technology because it helps them in their everyday lives and helps manage their jobs. However, technology should also be fun to use.

---

[1] The name 'ICQ' is a play on the phrase 'I seek you! It was one of the first Internet chat programs and also very popular.

Reachability and reliability is important, and they like tools that help them feel less stressed and more in control of their busy lives.

Lifestyle   Like the Young Pioneer, the Adult Pioneer lives a hectic life, often on the move. They spend more time away from home, for both private and professional reasons.

They are more likely than other segments to go to restaurants, meet friends and colleagues, and play and watch sports. They are also relatively active in their community and often act as opinion leaders.

They love experiencing new emotions and situations. Thus they are more likely to be interested in watching movies, both at home and in cinemas. They like listening to music, attending cultural activities such as the theater, concerts or museums and reading books to let their imaginations run free.

They enjoy nature, and have an interest in design and the arts; they are more interested in interior decorating and taking photos than other segments.

They like to keep up to date with what is happening in the world and read magazines about current affairs. When they have time to watch TV, they watch the news: current affairs and finance. Because they are interested in learning new things they also tend to watch documentary-style TV programs, including science and nature, history and travel.

Technology use   Pioneers enjoy cutting-edge technology and are eager to try it out in the most creative ways imaginable. They like the excitement and often find themselves addicted to new technology. Pioneers are heavy users of all kinds of technology and feel in control when using it. Mobile phones are important to them – without them they cannot keep up their fast-moving life.

They are often the first to explore and adopt new trends and this is true of technology. Adult Pioneers are sometimes even faster in their adoption of new technology than Young Pioneers, mainly because they can afford to fulfill their desires to a larger extent.

They are heavy users of mobile phones for both private and work reasons. They are also more exploratory when it comes to using their mobile phones and tend to use them for functions other than voice, for example synchronizing it with their computer or using it for accessing online services.
They are also heavy users of the Internet. They frequently browse for news and up-to-date information, including financial news. This group is most likely to use the Internet for banking/financial transactions.

A1.3   YOUNG MATERIALISTS (M1)

Demographic   15–24 years old, but especially 15–17 years old; slightly more males.

Singles, still live at home with their parents.

Still in school, many full-time students.

If they work they are more likely to work in shops.

Personality   Materialists seek pleasure and opportunity. They live in a world of amusement and leisure combined with moments of hard work to get wherever he or she wants to be. Appearances are very important to Materialists because they symbolize their rank and status in the groups they belong to. The clothes they are wearing, manifested by choosing the right brands and labels, become the uniform that announces they are 'members of the club'.

Young Materialists are devoted to friendship and fun, and live in a 'tribal' group society. They are more likely to have the newest mobile phones; they are rarely the ones who create an actual trend, but could turn a trend into a mass-market phenomenon.

They are cost conscious and do not always have the means to consume everything they would like to consume. They make up a large percentage of those who use prepaid, which they use to control their costs.

Lifestyle   The most important activity is socializing with friends – hanging out at discos and bars, listening to music, window shopping, and going to the movies or watching movies at home.

They have a high interest in activities that stimulate their need for entertainment, including playing computer/video games, reading cartoons and participating in sporting activities.

They watch more TV than Young Pioneers – entertainment programs, such as MTV, movies, sports, comedies, game shows, talk shows and reality TV.

They play sports and exercise to keep fit, but they eat whatever they like and are not willing to give up taste for a lower intake of calories.

Less frequent readers of newspapers than the average person, but tend to be heavier readers of magazines, especially those focusing on fashion, music celebrities and entertainment.

Technology use   Young Materialists are interested in new technology as entertainment; they find new technology exciting and it must be fun to use. They frequently play computer/video games, but also like to play games on mobile phones.

They possess a strong belief that technology increases opportunities to cultivate and create new social relationships, while being fun and convenient.

They use the Internet more than Adult Materialists but not as much as Pioneers. This group 'hangs out' on the Internet, making new friends over ICQ and chat rooms. They also acquire entertainment/leisure information, download music, send video clips and pictures as well as playing online games.

They are heavy SMS users because it is a cheaper way to keep in touch with other group members. They also find SMS fun to use.

Young Materialists personalize their mobile phones by downloading ring signals and pictures. This is a group that is truly interested in imaging and visual statements.

A1.4   ADULT MATERIALISTS (M2)

Demographic 25–39 years old; slightly more males.

More likely to be married or in a stable relationship.

Well educated with college degree.

Slightly higher household incomes.

Self-employed or work full-time in IT or sales; rarely work from home, but highly mobile.

Personality Adult Materialists place high value on a fulfilling career, as well as on knowledge and wisdom.

The image they present to others is important, as is looking good – many of them value being youthful.

Despite being older and more mature they tend to live for the moment and do not care too much about the consequences their actions of today might have tomorrow – now is the time to have fun and enjoy life.

Lifestyle They often have families and spend more time at home than Young Materialists. They like to entertain guests and show off their nicely decorated home. They are more likely to watch a video at home than go to the movies.

They love shopping and buy trendy clothes to show that they are up to date.

They like traveling for pleasure and enjoy taking photos to share the moment and show off about visiting a particular place.

This is the segment that is most likely to gamble on sports.

They read newspapers and magazines frequently; they prefer magazines that focus on fashion, sports, travel and celebrities and entertainment; they sometimes read business magazines to keep up to date with career opportunities.

The TV programs they prefer to watch are news, movies, sports, comedies and game shows.

Technology use   Materialists are 'early followers': they are quick to pick up on new trends but rarely start the trend themselves. They will often buy a product because they read about it in a trendy magazine or saw a person they admire (a movie star or someone at work or school) with the same product.

They like new technology and try to use it as much as they can, but it is also a way for them to show others that by mastering it they are up to date.

Their maturity is shown by their interest in new products that provide safety at home – they usually own property and want to protect it.

They are quite heavy users of the Internet, but not as heavy as Young Materialists. They use the Internet at work and home, where they look for entertainment/leisure information and exchange video clips and fun pictures with their friends. They are also more likely to use online banking, but are not particularly into ICQ/messaging or spending time in chat rooms.

They are more likely to have relatively new mobile phones – it is important to make a fashion statement by using a new model.

A1.5   EDUCATED SOCIABLES (S1)

Demographic   Most are 30–49 years old; both men and women.

Well educated – the best educated of all Take Five segments.

High household incomes.

Work full-time or are self-employed. More likely to have white-collars jobs or be teachers/trainers or specialists (doctor, lawyer, etc.). Many are managers. More likely to work in the public sector or in large organizations. They sometimes work from home.

Personality   Sociables are sophisticated with a mixture of rational thinking and emotional evaluation of things they encounter. They have strong values and beliefs and can often set their own desires aside in order to do something that is good for society. Many feel a strong responsibility towards the environment.

They are very avant-garde and flexible, but they are also cautious. They consider 'having fun' to be shallow. Sociables instead find pleasure in reading a good book, going for a swim or spending quality time with family.

They are not trend followers; they make up their own minds and base their decisions on rational lines of arguments. They are rarely the first to adopt new technology but can be quick to adopt once they see the advantages.

Lifestyle   This segment is highly mobile – they tend to spend slightly less time at home than Older Sociables, but still value their homes as an expression of who they are, which is why they like decorating it.

They like going to the movies or staying at home to watch movies with family and friends, taking it easy and perhaps having a glass of wine. They like exercising and playing sports.

They like an intellectual challenge and are interested in learning new things and discovering new cultures. They tend to be heavy readers of newspapers, but also read magazines on current affairs and traveling.

When they watch TV they like to watch the news and business updates, but also documentaries about other cultures, history or nature.

This is also one of the segments more likely to be involved in investing money in stocks and financial assets.

Technology use   Educated Sociables are early followers. They are likely to use services that are useful to them, with benefits such as saving time and becoming more efficient. They also expect technology to be user friendly.

They place high demands on their mobiles and are very cautious regarding the quality of the phone and the brand, which must come from a socially responsible supplier. They are interested in the esthetics of the phone, but for the pleasure of their own eyes rather than to show off to others.

They tend to use the Internet more to satisfy rational needs, and use it to search for information for work/school, find the latest news and stock quotes, and they are more likely to use it for banking.

Educated Sociables have not yet fully adopted new mobile services because they have not yet seen the full benefits of using them.

A1.6   OLDER SOCIABLES (S2)

Demographic
Most are 50–69 years old; more women.

Married/stable relationships and have children.

The second-lowest educated group; many have lower household incomes.

Many homemakers, unemployed or retired.

Work as trades people or have blue-collar jobs; very few managers. Work in smaller organizations and are more likely to work in shops. More likely to work from home.

Personality
Older Sociables tend to be a little more traditional and inward looking, focusing on themselves and their family in comparison with the Educated Sociable segment. They also place higher value on family, helpfulness, traditions and obedience than Educated Sociables.

Lifestyle
They tend to stay at home more and like doing gardening and looking after their homes to a greater extent.

Older Sociables like to do crafts, maybe as a way of expressing their own inner feelings.

They watch TV more often and also spend more time reading newspapers and magazines than younger Sociables, mainly because they have more time to do it.

They like exercising, taking care of their bodies to make them feel healthier, and they also enjoy being outdoors.

Technology use
They prefer to buy technology products that are made in their own country rather than abroad.

They find instruction manuals complex to use and think that too much computer use makes children less creative.

They feel more stressed by new technology but are generally not particularly scared of new technology, although some of them admit that it is a bit beyond them. Many more still think that you have to master it if you are to remain in control. This means they are generally late adopters rather than early followers.

Older Sociables have a lower usage and ownership of technical products and prefer to make calls to fixed-line phones than to mobile phones. They are more likely to say they do not need a mobile phone. They are also more skeptical about the impact mobile phones can have on health.

They are not heavy users of mobile phones today, and are more likely to have a prepaid subscription. When choosing a mobile phone operator, one trigger for them is the convenience of signing up. They are price-sensitive and more likely to be triggered by promotions and special offers, as well as recommendations by others.

A1.7   EDUCATED ACHIEVERS (A1)

Demographic   15–39 years old; more males.

More likely to be single.

Relatively well educated with a high school degree or higher.

Many are students.

Full-time employees working in IT or a production environment; less likely to telecommute.

Personality   Achievers are status seekers: status means having the 'right' job and being able to afford products and brands that they consider to be status markers.

They do not bother with petty details, and like to get straight to the point. They are good at finding what they want and work hard to get there. They trust themselves and are confident in their judgments, but they are not as decisive when it comes to purchasing products.

They are rich (or want to become rich) but are not very sophisticated in their tastes.

Lifestyle   Educated Achievers are a lot more adventurous than Older Achievers, mainly because they are younger and have higher incomes.

They are quite mobile and are usually out running errands, either private or for work, or meeting friends for a drink.

They wish they had more time to read books and attend cultural activities, but tend to end up on the sofa watching TV, listening to music, reading the newspaper, or playing computer/video games.

When they watch TV, they prefer to watch regular news programs as well as business news. They also like to watch soap operas to relax and dream about another life – the luxury life of most soap operas.

Educated Achievers like shopping, especially buying trendy items such as fashionable clothes.

They are more likely to take part in sports that meet their desires for status. They will, for example, exercise to look good or play golf to impress business partners. They are also more likely than other groups to bet on sports.

Technology use Up-to-date products are very important to Achievers. They show interest in new technological functions in mobile phones, but they are less interested in understanding *how* technology works and more inclined to show off new features rather than use them.

They consider new technology something that has to be mastered in order to show who is in control of it. They find new technology exciting and believe that it has the potential to make them more rational and efficient by, for example, always being reachable, being able to carry out online transactions (including betting), or listening to music wherever they are.

They are quite game-oriented and like to play games on their mobile phones and other devices. New mobile services and other 'cool' functions in mobile phones, such as Bluetooth, interest them, but only if they improve their image.

They consider the Internet an excellent tool for making them more efficient. They are, however, not very explorative Internet users and tend to use it mainly to find basic information for work/school, to browse, and to look up the latest news and information.

A1.8   OLDER ACHIEVERS (A2)

Demographic 30–69 years old, most are 55–69 years old; more females.

Married with children.

Low/basic education.

Lower household incomes.

Many homemakers, unemployed and retired.

Work in trades, blue collars jobs, restaurants or transport.

Personality   Older Achievers value material security more highly than younger, educated Achievers, mainly because they have lower household incomes and the status they are striving for is thus highly dependent on access to money.

They tend to worry more than others about economic recession, unemployment, inflation and not having enough money. However, they are not very socially responsible and tend to look after their own interests first without caring too much about the consequences of their actions.

Lifestyle   Older Achievers are more likely to stay at home than be out and about. Many would like to do more than they do today, but work or being at home caring for the family and keeping their homes neat takes all their time.

They like to socialize with friends and invite them to their tidy, nicely decorated homes. Because image is important to them, they also like gardening and try to keep their garden as neat as possible.

They wish they had more time to go window-shopping and buy trendy clothes for themselves.

They like taking photos occasionally, mainly to document family happenings, but also when there are guests visiting their home.

They sometimes read books and magazines – often romance or crime novels. They tend to read fashion, housekeeping and gossip magazines. They often check out the astrology pages in these magazines, but 'only for fun'.

They are usually so tired by the end of the day that they end up in front of the TV where they can dream about another, less harsh reality. Sometimes they listen to music and remember the good old days when they were young and had more freedom.

Technology use   They are less interested in new technology and are sometimes more skeptical than enthusiastic.

They often think new technology is a bit beyond them and instead prefer to continue using the technology they have already mastered. They find manuals for technological products complex and may even believe that computers make children less creative.

Almost half of them have never tried the Internet and do not wish to do so.

When it comes to mobile phones, low prices are key. Many will say that they do not need a mobile phone – this group has the lowest penetration of mobile phones on a global level.

If they have a mobile phone, they usually only make voice calls, but occasionally take the time to change their ringtones or send SMS messages.

A1.9   TRADITIONALISTS (T)

Demographic   Older, 40–69 years old, mostly 50–69 years old; slightly more women.

Married with kids.

Lower educational levels.

Lower household incomes.

More likely to be retired or homemakers.

If working, are more likely to have white-collar jobs in the government sector or in an institutional environment.

Personality   Traditionalists tend to look backwards. Today is as good as it gets – actually things were even better before – and they do not like change.

They tend to value history, roots and tradition more than other segments and are more likely to be nostalgic about how great things were once upon a time.

Caring for the environment is important. They also worry about things such as crime, drugs and disease, and would like to have more order in society. Justice, ethics and faith are also important. They are more likely to be religious and follow the traditional religion in their home country and often go to their place of worship to pray or meditate.

Because they do not like change, they tend to feel most comfortable in their own neighborhoods, where they have their friends. By staying in their immediate surroundings they do not have to interact with the unfamiliar.

Lifestyle   Traditionalists enjoy gardening, crafts and sometimes they like decorating their homes. They keep their cozy homes uncluttered and easy to clean. They sometimes invite friends over and enjoy socializing and remembering the good old days.

They watch slightly more TV than most people because it is an activity they can do at home. They like to watch the news, and are often horrified by the decline of society. They also enjoy watching sports, soap operas, dramas, mini-series, talk shows and comedies.

They are not heavy readers of newspapers or magazines, but when they do read magazines they tend to be about cooking or home issues, including parenting and decoration, but also about gossip and celebrities so that they can be horrified about the decline of morals.

Technology use

Traditionalists are late followers. They buy technology products only when they have become part of people's everyday lives. When they decide to buy a product they look for quality, mainly because once they have bought a product they will keep it for a long time.

They usually find new technology a bit scary and they are not frequent users of technology products. They find instruction manuals too complex and think that kids become less creative if they start using computers when they are too young.

A majority of them have never tried the Internet. Most do not wish to do so.

They agree that mobile phones are good for security reasons, but they still prefer to make calls to fixed phones. They worry about the health effects of mobile phones; despite this more than half of them are likely to own a mobile, but they have the oldest mobile phones of all segments and are not particularly interested in new services.

Appendix 2

The Most Mobile Work Roles

A2.1 TECH EMPLOYEES

Demographic	Young single males.
	Well educated.
Take Five	Early Adopters: many Pioneers and Young Materialists.
Employed in	Large companies, business services, telecoms, manufacturing, also banking/finance/insurance and utilities.
Typical roles	Computing, IT, technical control.
Mobile usage	Mobile on one location or within campus.
	They receive employer-provided tools.
Mobile services	Interested in all new mobile services.
	The most likely to do 'Pioneer' things like access corporate network while working remotely.

A2.2 SENIOR MANAGERS/KNOWLEDGE WORKERS

Demographic	Older married men with children.
	Well educated with high socio-economic levels.
Take Five	Slightly more likely to be Pioneers or Sociables.
Employed in	Many self-employed.
	Wholesale/retail, construction, and business services, but also arts/entertainment and banking/finance/insurance.

Mobile Media and Applications – From Concept to Cash: Successful Service Creation and Launch Christoffer Andersson, Daniel Freeman, Ian James, Andy Johnston, Staffan Ljung
© 2006 John Wiley & Sons, Ltd

Typical roles   Finance, marketing, human resources, consultants, doctors, lawyers, architects.

Mobile usage   Mobile around campus, town, national or international company and public sites.

Most likely to get employer-paid mobile phone bills, laptop, PDA.

Mobile services   Interested in most new mobile services.

A2.3   FIELD AND SERVICE STAFF

Demographic   The majority are males in the age-band between 25 and 39 years of age, married and living in a household with four to five family members. Average education.

Take Five   More likely to be Mature Materialists or Young/middle-aged Achievers

Employed in   Around a fifth work in the public sector and a large minority work for large companies with more than 200 employees.

Over-represented as full-time employees in industries of wholesale/retail, safety, and telecoms/utilities.

Typical roles   Salespeople (especially mobile salespersons), technical services (field sales), police, fire, military, etc.

Mobile usage   Receives mobile tools from employer, phone fixed in car/truck, and walkie-talkies/push-to-talk.

A small percentage get their mobile phone bills totally or partly financed by the employer.

Mobile services   They are interested in all new mobile services, but think it is especially important to constantly get a hold of the latest sport results and the financial market quotes.

A2.4   MOBILE INSTITUTIONAL

Demographic   Age 45–59, 50:50 male:female.

Average education.

Take Five   A combination of Pioneers, Sociables and Traditionalists.

Employed in   Many in large public sector institutions, mostly in health care and education.

Typical roles   Teachers, doctors, nurses, physical therapists, general administrative staff; often middle managers.

Mobile usage   Far less likely than Senior Managers to have employer paid mobile phone bills, laptop, or PDAs.

Mobile services   Interested in most new mobile services.

A2.5   SUPERMOBILE BLUE COLLAR

Demographic   Young males.

Low to middle education level.

Take Five   More likely to be Achievers and Young Materialists.

Employed in   Small companies including construction, manufacturing, travel/leisure/hotels/restaurants and utilities.

Typical roles   Trades people, mobile factory workers, store/shop/restaurant staff, service staff (cleaning, stocking shelves, etc.). Non-managers.

Mobile usage   They use a mobile phone for work but only a few get their bills paid by their employers. Few receive PDAs, pagers, or walkie-talkie-type phones.

Mobile services   Some interest in new services but driven more by cool devices – money buys new toys!

Index